Essentials of Toxicology

Essentials of Toxicology

Sreelakshmi Chakyar
Nitesh Agrawal
Tina Suresh

2026
Agri Horti Press
New Delhi-110 084

ISBN: 978-93-49690-34-9 (Hardbound)

Published By

Agri Horti Press

New Delhi

E-mail: agrihortipress@gmail.com

Digitally Printed at : Replika Press Pvt. Ltd.

PREFACE

Toxicology, the study of the adverse effects of chemical substances on living organisms, plays a critical role in understanding environmental, occupational, pharmaceutical, and food safety. With the continuous introduction of new chemicals and substances into the environment and the human body, the need for comprehensive toxicological knowledge has never been more pressing. From assessing the potential dangers of pollutants and pharmaceuticals to understanding the effects of toxins in our food supply, toxicology is at the heart of safeguarding public health and the environment.

This book, titled *Essentials of Toxicology*, offers a foundational overview of the field, designed to provide students, researchers, healthcare professionals, and policy makers with a clear and concise understanding of toxicological principles. Whether you are a beginner looking to enter the field or an experienced professional seeking to enhance your knowledge, this volume presents the key concepts and practices of toxicology in a comprehensive and accessible manner.

The book begins with an exploration of the basic principles of toxicology, including the mechanisms by which toxins interact with biological systems, how they are absorbed, distributed, metabolized, and excreted by the body, and the various ways toxins can cause harm. It explains key concepts such as dose-response relationships, toxicity testing, and the difference between acute and chronic toxicity. The book also delves into the role of toxicology in assessing risk, including the development of safety standards and regulations.

Essentials of Toxicology provides an in-depth examination of the various categories of toxins, such as environmental pollutants, industrial chemicals, pesticides, drugs, and naturally occurring toxins. It explores how toxicological research informs the regulation of these substances to protect human health and the environment. In addition, the book discusses the integration of toxicology with other scientific disciplines, such as pharmacology, environmental science, and epidemiology, to provide a more holistic understanding of toxin-related health risks.

Practical case studies and real-world examples are included throughout to highlight the application of toxicology in solving current global health challenges, from environmental pollution and food safety to the development of new pharmaceuticals. The book also emphasizes the ethical, legal, and social considerations surrounding toxicological research, particularly with respect to human and environmental safety.

It is our hope that Essentials of Toxicology will provide a comprehensive foundation for anyone involved in the study, research, and regulation of toxic substances, empowering readers to make informed decisions and contribute to the continued advancement of public health and safety.

Sreelakshmi Chakyar

Nitesh Agrawal

Tina Suresh

CONTENTS

1

INTRODUCTION

Toxicity is the degree to which a chemical substance or a particular mixture of substances can damage an organism. Toxicity can refer to the effect on a whole organism, such as an animal, bacterium, or plant, as well as the effect on a substructure of the organism, such as a cell (cytotoxicity) or an organ such as the liver (hepatotoxicity). By extension, the word may be metaphorically used to describe toxic effects on larger and more complex groups, such as the family unit or society at large. Sometimes the word is more or less synonymous with poisoning in everyday usage.

A central concept of toxicology is that the effects of a toxicant are dose-dependent; even water can lead to water intoxication when taken in too high a dose, whereas for even a very toxic substance such as snake venom there is a dose below which there is no detectable toxic effect. Toxicity is species-specific, making cross-species analysis problematic. Newer paradigms and metrics are evolving to bypass animal testing, while maintaining the concept of toxicity endpoints.

The degree to which a substance (a toxin or poison) can harm humans or animals. Acute toxicity involves harmful effects in an organism through a single or short-term exposure. Subchronic toxicity is the ability of a toxic substance to cause effects for more than one year but less than the lifetime of the exposed organism. Chronic toxicity is the ability of a substance or mixture of substances to cause harmful effects over an extended period, usually upon repeated or continuous exposure, sometimes lasting for the entire life of the exposed organism.

Toxicology is a branch of biology, chemistry, and medicine (more specifically pharmacology) concerned with the study of the adverse effects of chemicals on living organisms. It also studies the harmful effects of chemical, biological and physical agents in biological systems that establishes the extent of damage in living organisms. The relationship between dose and its effects on the exposed organism is of high significance in toxicology. Factors that influence chemical toxicity include the dosage (and whether it is acute or chronic); the route of exposure, the species, age, sex and environment.

HISTORY

Dioscorides, a Greek physician in the court of the Roman emperor Nero, made the first attempt to classify plants according to their toxic and therapeutic effect. Ibn Wahshiya wrote the Book on Poisons in the 9th or 10th century.

Mathieu Orfila is considered the modern father of toxicology, having given the subject its first formal treatment in 1813 in his Traité des poisons, also called Toxicologie générale.

In 1850, Jean Stas became the first person to successfully isolate plant poisons from human tissue. This allowed him to identify the use of nicotine as a poison in the famous Bocarmé murder case, providing the evidence needed to convict the Belgian Count Hippolyte Visart de Bocarmé of killing his brother-in-law.

Theophrastus Phillipus Auroleus Bombastus von Hohenheim (1493–1541) (also referred to as Paracelsus, from his belief that his studies were above or beyond the work of Celsus – a Roman physician from the first century) is also considered "the father" of toxicology. He is credited with the classic toxicology maxim, "Alle Dinge sind Gift und nichts ist ohne Gift; allein die Dosis macht, dass ein Ding kein Gift ist." which translates as: "All things are poison and nothing is without poison; only the dose makes a thing not a poison." This is often condensed to: "The dose makes the poison" or in Latin "Sola dosis facit venenum."

Basic Toxicology

The goal of toxicity assessment is to identify adverse effects of a substance. Adverse effects depend on two main factors: i) routes of exposure (oral, inhalation, or dermal) and ii) dose (duration and concentration of exposure). To explore dose, substances are tested in both acute and chronic models. Generally, different sets of experiments are conducted to determine whether a substance causes cancer and to examine other forms of toxicity.

Factors that influence chemical toxicity:

Dosage

(a) Both large single exposures (acute) and continuous small exposures (chronic) are studied.

Route of exposure

(a) Ingestion, inhalation or skin absorption

Other factors

(a) Species

(b) Age

(c) Sex

(d) Health

(e) Environment

(f) Individual characteristics

Testing Methods

Toxicity experiments may be conducted in vivo (using the whole animal) or in vitro (testing on isolated cells or tissues), or in silico (in a computer simulation).

Animal Testing Methods

The classic experimental tool of toxicology is animal testing. As of 2014, animal experiments provide information that is not available by other means about how substances function in a living organism.

Alternative Testing Methods

While testing in animal models remains the best method of estimating human effects, there are both ethical and technical concerns with animal testing.

Since the late 1950s, the field of toxicology has sought to reduce or eliminate animal testing under the rubric of "Three Rs" - reduce the number of experiments with animals to the minimum necessary; refine experiments to cause less suffering, and replace in vivo experiments with other types, or use more simple forms of life when possible. In some cases shifts away from animal studies has been mandated by law or regulation; the European Union (EU) prohibited use of animal testing for cosmetics in 2013.

Computer modelling is an example of alternative testing methods; using computer models of chemicals and proteins, structure-activity relationships can be determined, and chemical structures that are likely to bind to, and interfere with, proteins with essential functions, can be identified. This work requires expert knowledge in molecular modeling and statistics together with expert judgment in chemistry, biology and toxicology.

Dose Response Complexities

Most chemicals display a classic dose response curve – at a low dose (below a threshold), no effect is observed. Some show a phenomenon known as sufficient challenge – a small exposure produces animals that "grow more rapidly, have better general appearance and coat quality, have fewer tumors, and live longer than the control animals". A few chemicals have no well-defined safe level of exposure. These are treated with special care. Some chemicals are subject to bioaccumulation as they are stored in rather than being excreted from the body; these also receive special consideration.

Computational Toxicology

Computational toxicology discipline that develops mathematical and computer-based models to better understand and predict adverse health effects caused by chemicals, such as environmental pollutants and pharmaceuticals. Within the Toxicology in the 21st Century project, the best predictive models were identified to be Deep Neural Networks, Random Forest, and Support Vector Machines, which can reach the performance of in vitro experiments.

Toxicology as a Profession

A toxicologist is a scientist or medical personnel who specializes in the study of symptoms, mechanisms, treatments and detection of venoms and toxins; especially the poisoning of people. To work as a toxicologist one should obtain a degree in toxicology or a related degree like biology, chemistry or biochemistry. Toxicologists perform many different duties including research in the academic, nonprofit and industrial fields, product safety evaluation, consulting, public service and legal regulation.

Requirements

To work as a toxicologist one should obtain a degree in toxicology or a related degree like biology, chemistry or biochemistry. Bachelor's degree programs in toxicology cover the chemical makeup of toxins and their effects on biochemistry, physiology and ecology. After introductory life science courses are complete, students typically enroll in labs and apply toxicology principles to research and other studies. Advanced students delve into specific sectors, like the pharmaceutical industry or law enforcement, which apply methods of toxicology in their work. The Society of Toxicology (SOT) recommends that undergraduates in postsecondary schools that don't offer a bachelor's degree in toxicology consider attaining a degree in biology or chemistry. Additionally, the SOT advises aspiring toxicologists to take statistics and mathematics courses, as well as gain laboratory experience through lab courses, student research projects and internships.

Duties

Toxicologists perform many more duties including research in the academic, nonprofit and industrial fields, product safety evaluation, consulting, public service and legal regulation. In order to research and assess the effects of chemicals, toxicologists perform carefully designed studies and experiments. These experiments help identify the specific amount of a chemical that may cause harm and potential risks of being near or using products that contain certain chemicals. Research projects may range from assessing the effects of toxic pollutants on the environment to evaluating how the human immune system responds to chemical compounds within pharmaceutical drugs. While the basic duties of toxicologists are to determine the effects of chemicals on organisms and their surroundings, specific job duties may vary based on industry and employment. For example, forensic toxicologists may look for toxic substances in a crime scene, whereas aquatic toxicologists may analyze the toxicity level of wastewater.

Compensation

The salary for jobs in toxicology is dependent on several factors, including level of schooling, specialization, experience. The U.S. Bureau of Labor Statistics (BLS) notes that jobs for biological scientists, which generally include toxicologists, were expected to increase by 21% between 2008 and 2018. The BLS notes that this increase could be due to research and development growth in biotechnology, as well as budget increases for basic and medical research in biological science.

Toxicity

Toxicity is the degree to which a substance can damage an organism. Toxicity can refer to the effect on a whole organism, such as an animal, bacterium, or plant, as well as the effect on a substructure of the organism, such as a cell (cytotoxicity) or an organ such as the liver (hepatotoxicity). By extension, the word may be metaphorically used to describe toxic effects on larger and more complex groups, such as the family unit or society at large.

A central concept of toxicology is that effects are dose-dependent; even water can lead to water intoxication when taken in too high a dose, whereas for even a very toxic substance such as snake venom there is a dose below which there is no detectable toxic effect. Toxicity is species-specific, making cross-species analysis problematic. Newer paradigms and metrics are evolving to bypass animal testing, while maintaining the concept of toxicity endpoints.

TYPES OF TOXICITY

There are generally three types of toxic entities chemical, biological, and physical:

- Chemical toxicants include inorganic substances such as, lead, mercury, hydrofluoric acid, and chlorine gas, and organic compounds such as methyl alcohol, most medications, and poisons from living things. While some radioactive substances are also chemical toxicants, many are not: radiation poisoning results from exposure to the ionizing radiation produced by a radioactive substance rather than chemical interactions with the substance itself.
- Biological toxicants include bacteria and viruses that can induce disease in living organisms. Biological toxicity can be difficult to measure because the "threshold dose" may be a single organism. Theoretically one virus, bacterium or worm can reproduce to cause a serious infection. However, in a host with an intact immune system the inherent toxicity of the organism is balanced by the host's ability to fight back; the effective toxicity is then a combination of both parts of the relationship. A similar situation is also present with other types of toxic agents.
- Physical toxicants are substances that, due to their physical nature, interfere with biological processes. Examples include coal dust, asbestos fibers or finely divided silicon dioxide, all of which can ultimately be fatal if inhaled. Corrosive chemicals possess physical toxicity because they destroy tissues, but they're not directly poisonous unless they interfere directly with biological activity. Water can act as a physical toxicant if taken in extremely high doses because the concentration of vital ions decreases dramatically if there's too much water in the body. Asphyxiant gases can be considered physical toxicants because they act by displacing oxygen in the environment but they are inert, not toxic gases.

Measuring Toxicity

Toxicity can be measured by its effects on the target (organism, organ, tissue or cell). Because individuals typically have different levels of response to the same dose of a toxic substance, a population-level measure of toxicity is often used which relates the probabilities of an outcome for a given individual in a population. One such measure is the LD50. When such data does not exist, estimates are made by comparison to known similar toxic things, or to similar exposures in similar organisms. Then, "safety factors" are added to account for uncertainties in data and evaluation processes. For example, if a dose of a toxic substance is safe for a laboratory rat, one might assume that one tenth that dose would be safe for a human, allowing a safety factor of 10 to allow for interspecies differences between two mammals; if the data are from fish, one might use a factor of 100 to account for the greater difference between two chordate classes (fish and mammals). Similarly, an extra protection factor may be used for individuals believed to be more susceptible to toxic effects such as in pregnancy or with certain diseases. Or, a newly synthesized and previously unstudied chemical that is believed to be very similar in effect to another compound could be assigned an additional protection factor of 10 to account for possible differences in effects that are probably much smaller. Obviously, this approach is very approximate; but such protection factors are deliberately very conservative, and the method has been found to be useful in a deep variety of applications.

Assessing all aspects of the toxicity of cancer-causing agents involves additional issues, since it is not certain if there is a minimal effective dose for carcinogens, or whether the risk is just too small to see. In addition, it is possible that a single cell transformed into a cancer cell is all it takes to develop the full effect (the "one hit" theory).

It is more difficult to determine the toxicity of chemical mixtures than a pure chemical, because each component displays its own toxicity, and components may interact to produce enhanced or diminished effects. Common mixtures include gasoline, cigarette smoke, and industrial waste. Even more complex are situations with more than one type of toxic entity, such as the discharge from a malfunctioning sewage treatment plant, with both chemical and biological agents.

The preclinical toxicity testing on various biological systems reveals the species, organ- and dose- specific toxic effects of an investigational product. The toxicity of substances can be observed by:

- Studying the accidental exposures to a substance
- In vitro studies using cells/ cell lines
- In vivo exposure on experimental animals.

Toxicity tests are mostly used to examine specific adverse events or specific end points such as cancer, cardiotoxicity, and skin/eye irritation. Toxicity testing also helps calculate the No Observed Adverse Effect Level (NOAEL) dose and is helpful for clinical studies.

Global Classifications of Toxicity

For substances to be regulated and handled appropriately they must be properly classified and labelled. Classification is determined by approved testing measures or calculations and have determined cut off levels set by governments and scientists. While currently many countries have different regulations regarding the types of tests, amounts of tests and cut off levels, the implementation of Global Harmonization will begin unifying these countries as early as 2008.

Global Classification looks at three areas: Physical Hazards (explosions and pyrotechnics), Health Hazards and Environmental Hazards.

Health Hazards

The types of toxicities where substances may cause lethality to the entire body, lethality to specific organs, major/minor damage, or cause cancer. These are globally accepted definitions of what toxicity is. Anything falling outside of the definition cannot be classified as that type of toxicant.

Other Methods of Exposure and Severity

Skin corrosion and irritation are determined though a skin patch test analysis. This examines the severity of the damage done; when it is incurred and how long it remains; whether it is reversible and how many test subjects were affected.

Skin corrosion from a substance must penetrate through the epidermis into the dermis within four hours of application and must not reverse the damage within 14 days. Skin irritation shows damage less severe than corrosion if: the damage occurs within 72 hours of application; or for three consecutive days after application within a 14-day period; or causes inflammation which lasts for 14 days in two test subjects. Mild skin irritation minor damage (less severe than irritation) within 72 hours of application or for three consecutive days after application.

Serious eye damage involves tissue damage or degradation of vision which does not fully reverse in 21 days. Eye irritation involves changes to the eye which do fully reverse within 21 days.

Other categories of toxicity:

- Respiratory sensitizers cause breathing hypersensitivity when the substance is inhaled.
- A substance which is a skin sensitizer causes an allergic response from a dermal application.
- Carcinogens induce cancer, or increase the likelihood of cancer occurring.
- Reproductively toxic substances cause adverse effects in either sexual function or fertility to either a parent or the offspring.
- Specific-target organ toxins damage only specific organs.

- Aspiration hazards are solids or liquids which can cause damage through inhalation.

Environmental Hazards

Environmental hazards tend to focus on degradability, bioaccumulation and aquatic toxicity.

Mapping Environmental Hazards

There are many environmental health mapping tools. TOXMAP is a Geographic Information System (GIS) from the Division of Specialized Information Services of the United States National Library of Medicine (NLM) that uses maps of the United States to help users visually explore data from the United States Environmental Protection Agency's (EPA) Toxics Release Inventory and Superfund programs. TOXMAP is a resource funded by the US Federal Government. TOXMAP's chemical and environmental health information is taken from NLM's Toxicology Data Network (TOXNET) and PubMed, and from other authoritative sources.

FACTORS INFLUENCING TOXICITY

Toxicity of a substance can be affected by many different factors, such as the pathway of administration (whether the toxin is applied to the skin, ingested, inhaled, injected), the time of exposure (a brief encounter or long term), the number of exposures (a single dose or multiple doses over time), the physical form of the toxin (solid, liquid, gas), the genetic makeup of an individual, an individual's overall health, and many others. Several of the terms used to describe these factors have been included here.

Acute Exposure

A single exposure to a toxic substance which may result in severe biological harm or death; acute exposures are usually characterized as lasting no longer than a day.

Chronic Exposure

Continuous exposure to a toxin over an extended period of time, often measured in months or years; it can cause irreversible side effects.

Etymology

"Toxic" and similar words came from Greek "bow (weapon)" via "poisoned arrow", which came to be used for "poison" in scientific language, as the usual Classical Greek word for "poison" would transcribe as "io-", which is not distinctive enough. In some biological names, "toxo-" still means "bow", as in Toxodon = "bow-toothed" from the shape.

Principles and Procedures of Testing

The Acute Toxic Effects

Toxicology is concerned both with the nature and mechanisms of toxic lesions and the quantitative evaluation of the spectrum of biological changes produced by exposure to chemicals. Every chemical is toxic under certain conditions of exposure. An important corollary is that for every chemical there should be some exposure condition that is safe as regards man's health with the possible exception of chemical carcinogens and mutagens.

The quantitative evaluation of the biological changes caused by chemicals aims at the establishment of dose-effect and dose-response relationships that are of fundamental importance for health risk evaluation.

Defining Toxicity, Hazard, Risk, and Related Terms

In a general sense, the toxicity of a substance could be defined as the capacity to cause injury to a living organism. A highly toxic substance will damage an organism if administered in very small amounts; a substance of low toxicity will not produce an effect unless the amount is very large. Thus, toxicity cannot be defined without reference to the quantity of a substance administered or absorbed (dose), the way in which this quantity is administered (e.g. inhalation, ingestion, injection) and distributed in time (e.g. single dose, repeated doses), the type and severity of injury, and the time needed to produce that injury.

There is no generally agreed definition of "hazard" associated with a chemical, but the term is used to indicate the likelihood that a chemical will cause an adverse health effect (injury) under the conditions in which it is produced or used.

Risk is a statistical concept and has been defined by the Preparatory Committee of the United Nations Conference on the Human Environment, as the expected frequency of undesirable effects arising from exposure to a pollutant. Estimates of risk may be expressed in absolute terms or in relative terms. The absolute risk is the excess risk due to exposure. The relative risk is the ratio between the risk in the exposed population and the risk in the unexposed population.

Safety is a term that has been used extensively but is difficult to define. One definition is that "safety" is the practical certainty that injury will not result from the substance when used in the quantity and in the manner proposed for its use.

This definition is of little use unless "practical certainty" is defined in some way, for example, in terms of a numerically specified low risk. Another view is that "safety" should be judged in terms of socially "acceptable" risks. Such judgments are largely outside the scope of scientific evaluation but nevertheless require assessment both of the probabilities of various adverse effects and of their severity in terms of human health or other concerns.

Laboratory Testing

Human data on the toxicity of chemicals are obviously more relevant to safety evaluation than those obtained from the exposure of experimental animals. However, controlled exposures of man to hazardous or potentially hazardous substances are limited by ethical considerations and information obtained by clinical or epidemiological methods must be relied on. Where such information is not available, as in the case of all new synthetic chemicals, data must be obtained from tests on experimental animals and other laboratory procedures. The degree of confidence with which human health risks can be estimated from laboratory data depends on the quality of the data, and the selection of appropriate laboratory testing procedures is the main subject of this monograph.

Toxicological Field Studies

In the laboratory, only a small number of animal species are available for testing. The testing of wild species, living in cages under field conditions, may be useful but sometimes presents a variety of problems. Successful trials require a large enough site (about 8 ha; 20 acres) with adequate and varied populations of birds, mammals, fish, insects, and other species, and the area studied must be considerably greater than that treated. Data obtained from field trials of chemicals are of considerable value in supplementing data obtained with laboratory animal species and in validating the projection of experimental results to the ecosystem, including man. Studies of random events in natural ecosystems can also provide useful data.

Sensitive analytical techniques now make it relatively simple to conduct field studies in man by monitoring levels of a chemical or its metabolites in blood, urine, hair, or saliva; this biological monitoring together with environmental monitoring provides important information on the exposure of man. Regular periodic determination of the profile of certain plasma enzymes and other biochemical variables in the subject provides another valuable method for monitoring health effects particularly under occupational exposure conditions; changes in these profiles may provide early warning of damage by toxic chemicals.

Ecotoxicology

A new subdivision of toxicology, "ecotoxicology", has emerged following observations that some persistent chemicals can exert toxic effects at several points in an ecosystem. The appearance of a chemical or the manifestation of a toxic effect may occur far away from its initial point of introduction into the environment. Methods for assessing the extent and significance of the movement of pollutants and their degradation products through the environment to target systems are discussed in a recent publication.

PRIORITIES IN THE SELECTION OF CHEMICALS FOR TESTING

In principle, all new chemicals require safety evaluation before manufacture and sale, but, because of the large number of chemicals that represent a possible hazard to human health and limited resources, it is necessary to give priority to those that are directly consumed by man, such as drugs and food additives, and those that are widely used such as pesticides or household consumer products. Industrial chemicals that can escape into the working or general environment or can contaminate other products are another category of concern.

Compounds of suspected high acute, chronic, or delayed toxicity (such as carcinogenicity) or of high persistence in the environment, or compounds which contain chemical groups known to be associated with these properties, deserve the highest priority. This also applies to compounds known to inhibit metabolic deactivation of chemicals as they may represent a more insidious form of toxicity.

Chemicals resistant to metabolism, especially metabolism by microflora, will have a high environmental persistence. Many halogenated compounds come into this category, and should, therefore, have some degree of priority. Compounds that accumulate in food chains or are stored in the body, e.g. methylmercury and DDT, will be a matter of concern. Such compounds are often highly lipid-soluble or strongly bound to tissue proteins, or may undergo enterohepatic recirculation with consequent slow excretion resulting in accumulation in the organism.

Physicochemical properties can be an important consideration in setting priorities for testing potential environmental pollutants. For example, biomagnification of stable, fat-soluble substances may lead to contamination of human food supplies as well as to adverse effects in wildlife at the higher levels of food chains, even though the intended use and sites of application of the substance would suggest that primary exposure of these species is unlikely. Physicochemical properties such as vapour pressure, and particle size and density are important in predicting the atmospheric transport of chemicals. Adsorption of a chemical on soil particles may increase the likelihood that the material will become airborne or be transported by watercourses and subsequently deposited in areas remote from its site of application, or it may retard the movement of a chemical through ground water and thus reduce the likelihood of contamination of ground water supplies near the site of application.

Even though certain predictions and comparisons of environmental distribution and biomagnification of chemicals in the environment may be made theoretically on the basis of the physicochemical properties of the substances in question, more definitive information of this nature can be obtained experimentally by the use of model ecosystems such as those described by the researchers. These model ecosystems may be oversimplified, and they should not replace experimental field studies or programmes for monitoring environmental contaminants. However, their use in an early phase of the overall evaluation of the toxicity of environmental chemicals may:

- Help to determine order of priority of chemicals for study,
- Identify the components of the environment (food, water, air) most likely to be a source of human exposure and
- Suggest whether the chemicals are likely to accumulate in human tissues.

Furthermore, the systematic application of such model systems to structure-distribution studies may help in determining with greater certainty those physicochemical properties of substances that are most useful in predicting the distribution and effects of chemicals in ecosystems.

Information on production, use, and disposal are of great importance in determining the sources and quantities of a chemical released into the environment, in assessing the possible extent of human exposure, and in identifying human populations that are likely to be exposed.

In conclusion, essential criteria for priority in the selection of chemicals for testing are:

- Indication or suspicion of hazard to human health and type and severity of potential health effects;
- Probable extent of production and use;
- Potential for persistence in the environment;
- Potential for accumulation in biota and in the environment, and

- Type and size of populations likely to be exposed. A chemical of first priority for testing would rate highly with respect to all or most of these criteria.

The Extent of Toxicity Testing Required

The extent of the toxicity testing required will depend on a variety of considerations, and generally valid procedures cannot be proposed. As a first step, it may be useful to make an approximate estimation of toxicity based on the chemical structure and the physical and chemical properties of the substance, and on known correlations of these variables with biological activity. These considerations may be of value for decisions on safety measures to be taken during initial laboratory work. Extrapolation and interpolation in homologous series may also be of value for decisions on safety measures to be taken during initial laboratory work, but for some series of chemicals this is not applicable.

A preliminary evaluation of toxicity should start when chemicals are synthesized in the laboratory stage of the development of an industrial process. The full evaluation of the chemicals involved, both in respect to occupational and general population exposure, and assessment of possible air, water, and food contamination, should be initiated later, when it has been decided to proceed with full-scale production of the chemical. Toxicity data obtained during the development stages of a technological process could provide information concerning the health hazards not only of the raw materials and products, but also of the various other substances used or produced as intermediates in the technological process, and of gaseous and other wastes. Toxicological evaluation may also help in the selection of an alternative technological process, less hazardous to health.

Waste disposal by dispersion in air and water, the ease of environmental degradation of the chemical, and the toxicity of the degradation products, are other problems that need attention at an early stage in the toxicological evaluation of new chemicals. For example, resistance to degradation has to be taken into account when formulating health criteria regulating the application and disposal of pesticides.

This phasing of toxicological studies may be useful in coordinating testing at national and international levels.

Environmental and health standards will need to be defined preferentially for those chemicals that show a significant degree of toxicity and represent a health hazard, and are likely to be used widely in industry, agriculture, or in consumer products.

Changes and developments in industrial processes, the development of new chemicals, and changes in the use of existing chemicals, may lead to new or increased hazards. This calls for a continuous re-evaluation of priorities.

DOSE-EFFECT AND DOSE-RESPONSE RELATIONSHIPS

Effect and Response

"Effect" and "response" are often used interchangeably to denote a biological change, either in an individual or in a population, associated with an exposure or dose. Some toxicologists have, however, found it useful to differentiate between an effect and a response by applying the term "effect" to a biological change, and the term "response" to the proportion of a population that demonstrates a defined effect.

In this terminology, response means the incidence rate of an effect. For example, the LD50 value may be described as the dose expected to cause a 50% response in a population tested for the lethal effect of a chemical. This distinction will be made in the present monograph, although it should be recognized that this terminology is not generally accepted.

An effect can usually be measured on a graded scale of intensity or severity and its magnitude related directly to the dose. Certain effects, however, permit no gradation and can be expressed only as "occurring" or "not occurring". Such effects are usually called "quantal". Typical examples of quantal effects are death or occurrence of a tumour.

The toxic action of chemicals usually affects the whole organism but the primary damage may be localized in a specific target organ or organs in which the toxic injury may manifest itself in terms of dysfunction or overt disease. According to Sanockij, the specificity of acute toxic action can be expressed in terms of a "zone of specific action" (Zsp) which is the ratio between the threshold dose of an acute effect at the level of the total organism and the threshold dose for an acute effect at a specific organ or system. If $Zsp > 1$, the toxic action is specific; if $Zsp < 1$, it is non-specific.

Acute effects are those that occur or develop rapidly after a single administration but acute effects may appear after repeated or prolonged exposure as well. Chronic effects may also result from a single exposure but more often they are a consequence of repeated or prolonged exposures. Chronic effects are characterized not only by their duration but also by certain pathological features. They may arise from the accumulation of a toxic substance or its metabolites in the body, or from a summation of acute effects. The latent period (or the "time-to-occurrence" of an observable effect) may sometimes be very long, particularly if the dose or exposure is low.

Dose-effect and Dose-response Curves

Dose-effect curves demonstrate the relation between dose and the magnitude of a graded effect, either in an individual or in a population. Such curves may have a variety of forms. Within a given dose range they may be linear but more often they are not. Research has discussed various transformations that can be used to make dose-effect curves linear.

Dose-response curves demonstrate the relation between dose and the proportion of individuals responding with a quantal effect. In general, dose-response curves are S-shaped (increasing), and they have upper and lower asymptotes, usually but not always 100 and 0%. One way of explaining the shape of dose-response curves is that each individual in a population has a unique "tolerance" and requires a certain dose before responding with an effect. There exists, in principle, a low dose to which none will respond and a high dose to which all will respond.

For each effect there will usually be a different dose-response curve. Researchers have discussed the relationship between dose, effect, and response and its graphical representation in a three-dimensional model.

If the experiment or observation is well designed, the dose-response relationship will be based on data from many individuals over a range of doses from minimum to maximum response. Mathematical and statistical procedures are then used to establish the curvilinear relationship that provides the best fit to all of the data, expressed as mean values with their standard deviations at different doses.

It should be pointed out that the shape of the dose-response curve for the same substance and the same animal species may vary with changes in experimental conditions, such as changes in the way in which the dose is distributed in time.

In evaluating human exposure to environmental chemicals, the dose will usually be estimated as a function of concentration and time. In some cases the concentration will be fairly constant and then the time-effect and time-response relationships will be similar to the dose-effect and dose-response relationships. However, in many cases the concentration will vary, as will the time of exposure to specific concentrations, and integrated relationships of dose-concentration- time must be considered as well as dose-effect and time-effect relationships.

Haber's rule (ct=k) states that the product of concentration (c) and time (t) results in a constant intensity of effect (k) for some gases. This formula was later changed to ctb = k (where b is constant) which fitted other biological data better, although it also has its limitations. The extrapolation of concentration-time relationships has been used successfully to obtain predictions of response following long-term inhalation exposure to low concentrations.

Concentration-time relationships, such as the variation of the fraction of the dose with time as in combinations of short-term peak concentrations and prolonged low-level concentrations in air pollution, and variable cycles of exposure, may influence the toxic effect. Few systematic attempts to evaluate these factors have been made, although have described some principles for setting air quality standards from this viewpoint, and Pinigin (1974) has dealt with the problems of intermittent inhalation exposure.

Toxic Effects due to a Combination of Chemicals

When an organism is exposed to two or more chemicals, their joint action may be:

- Independent — when the chemicals produce different effects or have different modes of action;
- Additive — when the magnitude of an effect or response produced by two or more chemicals is numerically equal to the sum of the effects or responses that the chemicals would produce individually;
- More than additive — often called potentiation or synergism;
- Less than additive (antagonism, inhibition).

More specific terminology may be used when the mechanisms of joint action are known or when definite assumptions are made about them. The time intervals and sequences between exposures to different chemicals are extremely important, and the quality as well as the degree of joint action may depend on these variables. Furthermore, the joint action at lethal dose levels may be quite different from that at low dose levels, when the effects or responses are often only additive or independent.

Most statistical models for joint action have been developed for situations in which two or more chemicals are administered simultaneously or within a short (few minutes) time interval. A model proposed by Finney is often used for predicting the acute joint toxicity of chemicals. The model is strictly applicable to mixtures of chemicals that act at the same site, producing the same type of acute toxic effect and having parallel regression lines of probits against log doses. For a mixture of, for example, three chemicals, the equation for the median effective dose (ED50) is (eq.) where contour integralA, contour integralB and contour integralC are the fractions of substances A, B, and C in the mixture. When all the values on the right hand side of equation (1) are known, a predicted ED50 (assuming additive joint action) can be calculated and compared with the actual ED50 of the mixture determined experimentally. A smaller than predicted ED50 demonstrates a more than additive response (synergism), a greater than predicted ED50 indicates a less than additive response (antagonism). Researchers demonstrated that this equation can give satisfactory results under conditions that are less restrictive than stated above, for example in identifying the type of acute joint action among randomly selected industrial chemicals. Researcher applied the equation to the estimation of maximum allowable concentrations for occupational exposure to mixtures of substances that exercise a "similar joint action", e.g. benzene and toluene. Another model for estimating the results of joint action has been developed using the isoeffective concentrations instead of ED50 in equation (1).

The possibility of predicting the type of joint action is enhanced if there is information on the metabolism and disposition of the chemicals. Basic principles concerning the kinetics of reactions of chemicals with primary sites (tissue receptor sites) and with secondary sites are important in considering the joint action of chemicals. The relevant factors seem to be the relative affinities at the sites of action (e.g. target enzymes, neuroeffector sites, and other vital target sites), and at the sites of loss or sinks (e.g. detoxifying enzymes, nonvital tissue binding sites, pathways of excretion, and storage sites), and the intrinsic activity at the sites of action. Since there is a limited number of sites of action and sinks within any organism, there will be a limited dose range within which synergism or antagonism can be demonstrated. This, of course, is only one area where more information could help in predicting the effects of the joint action of chemicals. Other areas where knowledge is insufficient are the possible effects of low-level, prolonged exposures to mixtures of chemicals and the effects of multiple stresses including chemicals, physical factors such as heat and noise, and pre-existing disease.

Interpretation of Laboratory Data

It is essential that all experiments to evaluate toxicity should be designed to be scientifically meaningful, and should not be conducted merely to comply with statutory regulations. Thus, the evaluation of each new chemical will not be an identical task and procedures will differ, to some extent, from one compound to another. The protocol for an experiment will evolve gradually during the experiment, in accordance with earlier findings. It is useful to have laboratory data validated by a study of the mechanisms involved in the development of the toxic lesion. Furthermore, numerous endogenous and environmental variables can modify the toxicity of chemicals, as discussed in subsequent chapters. In some instances, the influence of these variables is known, and can be controlled, but often this is not the case and this may cause serious difficulties in the interpretation of laboratory toxicity data.

In the present context, we are mainly interested in the interpretation of laboratory data with a view to their application in the evaluation of the health risk to man. The discussion will therefore be limited to a few topics that are particularly relevant in this respect.

DISTINCTION BETWEEN ADVERSE AND NON-ADVERSE EFFECTS

An adverse, or "abnormal" effect has often been defined in terms of a measurement that is outside the "normal" range. The "normal" range, in turn, is usually defined on the basis of measured values observed in a group of presumably healthy individuals, and expressed in statistical terms of a range representing 95% confidence limits of the mean or, for individuals, in terms of 95% "tolerance" limits established with a derived degree of confidence (95% or 99%). An individual with a measured value outside this range may be either "abnormal" in fact, or one of that small group of "normal" individuals who have extreme values. The distinction between "normal" and "abnormal" values based on statistical considerations may be used as a criterion for adverse effects, if the exposed population consists of adult, generally healthy individuals, subject to periodical medical examination, such as workers. Departures from "normal" values associated with a given exposure will then be considered as adverse effects, if the observed changes are:

- Statistically significant ($P < 0.05$) in comparison with a control group, and outside the limits ($m \pm 2$ s) of generally accepted "normal" values;
- Statistically significant ($P < 0.05$) in comparison with a control group, but within the range of generally accepted normal values, provided such changes persist for a considerable time after the cessation of exposure; and
- Statistically significant ($P < 0.05$) in comparison with a control group, but within the "normal" range, provided statistically significant departures from the generally accepted "normal" values become manifest under functional or biochemical stress.

This statistical definition of adverse effects is less suitable for the general population which includes some groups that may be specially sensitive to environmental factors, particularly the very young, the very old, those affected with disease, and those exposed to other toxic materials or stresses. In this case, it is practically impossible to define "normal" values, and any observable biological change may be considered as an adverse effect under some circumstances. For this reason, attempts have been made to set criteria for adverse effects based on biological considerations and not only on statistically significant differences with respect to an unexposed population (control group). Although there is no general agreement on such criteria and the ultimate decision on what is an adverse effect will have to depend, in each case, on experience and expert judgment, it may nevertheless be useful to give examples of such criteria, which illustrate at the same time how different such criteria may be.

A Committee for the Working Conference on Principles of Protocols for Evaluating Chemicals in the Environment defined nonadverse effects as the absence of changes in morphology, growth, development, and life span. Furthermore, nonadverse effects do not result in impairment of the capacity to compensate for additional stress. They are reversible following cessation of exposure without detectable impairment of the ability of the organism to maintain homeostasis, and do not enhance susceptibility to the deleterious effects of other environmental influences.

On the other hand, adverse effects may be deduced as changes that:

- "Occur with intermittent or continued exposure and that result in impairment of functional capacity (as determined by anatomical, physiological, and biochemical or behavioural parameters) or in a decrement of the ability to compensate additional stress;
- Are irreversible during exposure or following cessation of exposure if such changes cause detectable decrements in the ability of the organism to maintain homeostasis; and
- Enhance the susceptibility of the organism to the deleterious effects of other environmental influences."

Soviet toxicologists emphasize that criteria for differentiating between adverse and nonadverse effects should not be based on overt pathology (e.g. inflammation, necrosis, hyperplasia), and have proposed, inter alia, a number of criteria based on metabolic and biochemical changes. Such changes are considered to be adverse if:

- The metabolism of a substance becomes less efficient or the elimination of a substance (expressed in terms of biological half-time, T′) slows down with increasing doses of the substance;
- Enzymes that have a key significance in metabolism are inhibited;
- The inhibition of a certain enzyme results in an increase in the concentration of the corresponding natural substrate in the body and/or in a decreased

capacity to metabolize the specific substrates in a loading test;

- The relative activities of different enzyme systems are changed (e.g. the ratio of the activities of asparagine and alanine transaminases.

Researcher also attaches great importance to the changes in the pattern of isoenzymes in the blood, and to the changes in the subcellular membranes (e.g. lysosomal membranes) resulting from the action of toxic substances.

Differentiation between "nonadverse" and "adverse" effectsrequires considerable knowledge of the importance of reversiblechanges and subtle departures from "normal" physiology and morphologyin terms of the organism's overall economy of life, ability to adaptto other stresses, and their possible effects on life span. Newer and improved methods of research have increasingly provided more sensitive tests for subtle biological deviations such as induction of enzymes of the smooth endoplasmic reticulum of the liver, or reversible hypertrophy of the liver. These types of changes are produced by relatively low doses of many chemicals and they are considered by some authors to be adaptive and generally useful to health, and by others to be indicative of injury. One of the most challenging areas for basic research in toxicology today is the acquisition of data that can be used to estimate whether, or under what conditions, subtle changes in enzyme activities, nerve action potentials, altered behavioural reaction etc. indicate impairment of physiological function or predict impending development of more serious irreversible injury, should exposure to the chemical continue.

In addition to all these considerations, the possibility must be kept in mind that an effect may not be seen because the number of animals studied was inadequate, the observation time was too short, or for other reasons.

Threshold: Practical and Theoretical Considerations

The concept of "threshold" is complex and the term has to be carefully defined, so that statements concerning this concept in relation to the protection of human health are not confused by semantic differences. A distinction should be made between the threshold for individuals and thresholds for limited groups of individuals or general populations.

The dependence of effect or response on the dose of a chemical has already been discussed. As a rule, the intensity of the effect or response decreases with reduction in dose, and a biological reaction often reaches zero before the dose becomes equal to zero. Below a certain limiting exposure level, or dose, i.e. below the threshold, a chemical substance may not elicit a toxic effect. The threshold for an adverse effect of a chemical is defined by some toxicologists as the minimum exposure level or dose that gives rise to biological changes beyond the limits of homeostatic adaptation. True homeostatic adaptation should be carefully distinguished from pathological processes.

The existence of a threshold for all adverse effects is, however, still a matter for discussion. Sanockij has provided data which show that small quantities of environmental chemicals may not reach their receptor because the rate of elimination or metabolic degradation is relatively more effective with smaller doses. It has also been suggested that where effective repair processes are present, even if a substance interacts with the receptor, it need not necessarily produce an adverse effect.

For some toxic effects, such as neoplastic disease or mutations of genetic material, it has been assumed by some authors that a single molecule of a chemical is sufficient to initiate a process that may progressively lead to an observed, harmful effect. For this reason, it may not be possible to demonstrate that a threshold dose for a carcinogen or a mutagen exists.

Other scientists view carcinogenic or mutagenic chemicals as toxic entities that may have special properties with regard to the nature and characteristics of their adverse effects, but are subject to the same physicochemical and biological interactions that are considered to result in a threshold dose for other chemicals.

The question of the existence of a threshold for carcinogens and mutagens was recently discussed by a WHO Scientific Group, which concluded that "the existence of a threshold may be envisaged. Nevertheless, the difficulties of determining a threshold for a population are great. Therefore, mathematically derived conclusions that it is impossible to demonstrate no-effect levels experimentally cannot be ignored". A "no-effect" level for a group of animals may occur because the dose is really below the theoretical no-effect level (i.e. below the threshold) or because the number of animals is too small. For example, in an experiment with 20 animals, it is possible that none of the animals will show an effect whereas in an experiment with 100 animals some response might be seen. However, an upper limit for the probable response can be estimated statistically. For instance, if in an experiment with 100 animals, no response has been observed, it can be shown that there is a 95% probability that, under the conditions of the experiment, the upper limit of response is 3%, and that there is a 99% probability that the response will not exceed 4.5%. Even in an experiment with 1000 animals showing no response, the upper 95% confidence limit of response is 3 animals showing an effect per 1000 treated animals.

Another reason for not having seen a response in an experiment may be that the time of observation was too short. This may be the case, for example, when the quantal effect considered is a cancer, with a long latent period between exposure and appearance of tumours.

For these reasons the "no-effect level" has no real meaning and a better term is "no-observed-effect level".

EXTRAPOLATION OF ANIMAL DATA TO MAN

In many cases, studies with laboratory animals make it possible to predict the toxic effects of chemicals in man. However, it is important to realize that experimental animal models have their limitations, and that the accuracy and reliability of a quantitative prediction of toxicity in man depend on a number of conditions, such as choice of animal species, design of the experiments, and methods of extrapolation of animal data to man.

Researchers considered the criteria for the adequacy of an experiment to be used for the extrapolation of animal data to man. They include: test animal species and strain (the animal should be susceptible to induction of the effects under consideration); the number of animals; the route of administration (which should include the routes of human exposure); and the physical state and chemical form of the agent. The side effects of the chemical and its organ specificity should also be taken into account in the design of the experiment. In interpreting the results, attention should be paid to adequate survival of the animals, to possible intercurrent disease, the quality and extent of pathological data, the quality and extent of relevant data collection during the experiment, and the availability of data at the time of interpretation.

Species Differences and Related Factors

The most difficult problem in the extrapolation of animal data to man is the conversion from one species to another. For most substances, the pathogenesis of poisoning is the same in man and other mammals, and for this reason the signs of intoxication are also analogous. Thus, quantitative rather than qualitative differences in toxic response are most common. Man may be more sensitive than certain laboratory animals but there are also many cases where some animal species are more sensitive than man. For example, the mouse is most sensitive to atropine, the cat is less sensitive, while the dog and the rabbit tolerate atropine in doses 100 times higher than the lethal dose for man. However, the dog is more sensitive to hydrocyanic acid than man.

Species differences in sensitivity can often be explained by differences in metabolism, in particular by quantitative and qualitative differences in the ability of an enzyme to detoxify chemicals, and also by differences in the rates of absorption, transport, distribution, and elimination of chemicals Researchers discussed various factors to be considered in the selection of animal models for pharmacotherapeutic studies in relation to the steps that intervene between administration of the drug (or chemical) and the arrival of the compound at the ultimate sites of action. After oral administration, absorption in standard laboratory animals is generally considered to be very similar to man, although there are quantitative differences for some compounds. For example, species differences in the absorption and action of some compounds are related to differences in the bacterial flora of the gastrointestinal tract. Rall further concluded that the distribution and storage of drugs are reasonably consistent in mammalian species, including man, although plasma binding tends to be more extensive in man than in small mammalian species. Urinary excretion in different animal species depends to some extent, on their different diets, since diet influences urinary pH and thus the extent of ionization of compounds. Biliary excretion is quite variable from species to species and apparently is more extensive in mice and rabbits than in rats or man. Species differences in response to chemicals appear to be mainly related to rates of biotransformation which are generally more rapid in small laboratory animals than in man.

One of the most potent bladder carcinogens, 2-naphthalenamine (2-naphthylamine), produces bladder cancer in the dog, hamster, and man, but not in the rat, rabbit, or guineapig. Species differences in the carcinogenicity of 2-fluorenylacetamide (2-acetaminofluorene) have been attributed to the different extents of metabolism to the proximate carcinogen, the N-hydroxy derivative. Similarly, strain differences in metabolism may also affect toxicity.

If metabolic information is available, differences in absorption, distribution, biotransformation, and elimination of toxic substances in man and animals should be taken into account when selecting experimental animals.

Species differences in toxicity may also be due to differences in cellular transport. Aflatoxin, which is more toxic to rats than to mice, both as an acute poison and as a carcinogen, is transported more slowly into the liver cells and is metabolized more rapidly in the mouse than in the rat.

In determining the required duration of an animal experiment, it is often useful to compare the life span of the animal with that of man. Using the "body weight rule", the average life span for 70 species of mammals showed a linear correlation with body weight, but the average life span of man was found to be an exception. The regression equation obtained from a study of many mammals showed that the average life span for a mammalian representative, having the same body weight as man (70 kg) was equal to 15 years.

Thus, if this assumption is accepted the average life span of a rat (about 2.5 years) corresponds to only 15-17 years of a man's life. This inconsistency in the life spans of man and experimental animals should be taken into account in the design and interpretation of animal experiments for the evaluation of toxicity to man.

There are other problems in the evaluation of toxicity to man from experiments on animals, such as where an effect is difficult to measure or where similar conditions are difficult to obtain in animal models, for example, intelligence and the more esoteric behavioural changes. Furthermore, in animal experiments, the effects of social factors, so important to man, cannot be evaluated.

For these reasons, when extrapolating from animals to man it is prudent to apply a species conversion factor which should be determined on the basis of biological considerations and the available information on the test species. There is no definite rule for the species conversion factor. If the extrapolation of data is based on the most sensitive species tested, some toxicologists use a factor of 1, but others recommend a factor as large as 10.

The unit of dose to be used has also to be considered in the extrapolation of data to man and it has recently been recommended that the dose per unit surface area approximately equivalent to the weight raised to the power 2/3 should be used. If the dose is given in terms of dietary concentration, there seems to be no need to make the surface area adjustment.

A separate problem, to which there appears to be no satisfactory answer at present, is the conversion from an inbred animal strain to a genetically highly heterogeneous human population.

Safety Factors

In almost all instances, laboratory data on the toxicity of chemicals are drawn from experiments in which the adverse effect occurs at a considerably higher incidence rate than would be acceptable in man. For this reason alone, and apart from the biological differences between laboratory species and man, an extrapolation from a known dose-response range to an unknown range is necessary. Indeed, essentially the same problem arises when a human accident or epidemiological data are used as the starting point.

Traditionally, a safety factor has been introduced to provide for uncertainties in extrapolation from animals to man, and from a small group of individuals to a large population. Such safety factors have ranged from 1 to as much as 5000. Because of the current uncertainty regarding the mathematical and biological reliability of methods for extrapolating from high doses to low doses, primary dependence on somewhat arbitrary safety factors continues. However, means of extrapolating from high to low doses are being intensively studied at the present time, especially with respect to carcinogenicity.

Most regulatory authorities rely on the use of safety factors but there are no precise guidelines for deciding the appropriate size of such a factor. Researchers have discussed the rationale for different safety factors. In general, the size of the safety factor will depend on (a) the nature of the toxic effects, (b) the size and type of population to be protected, and (c) the quality of toxicological information available. A factor of 2 to 5 or less may be considered as sufficient if the effect against which individuals or a population are to be protected is not regarded as very severe, if only a small number of workers are likely to be exposed, and if the toxicological information is derived from human data. On the other hand, a safety factor as large as 1000 or more may be required if the possible effect is very serious, if the general population is to be protected, and if the toxicological data are derived from limited experiments on laboratory animals. In some cases, the safety factor may be a value that has been used with reasonable success and is, therefore, perpetuated.

For most food additives that are not considered to be carcinogenic, it has been the accepted practice to divide the no-adverse-effect dose (i.e., the maximum ineffective dose) in animals by 100, to arrive at an acceptable daily intake (ADI) for man. For pesticides and certain environmental chemicals, safety factors ranging from less than 100 to several thousand have been used. For some occupational exposures, and for certain air pollutants much smaller safety factor have been proposed in the range of 2-5. Safety factors have also been proposed for carcinogenic chemicals ranging from 100 to about 5000 (Weil, 1972a) but they have not been generally accepted.

Low-dose Extrapolation

Low-dose extrapolation is based on mathematical models that are used to predict the response at a given low dose or to predict that dose which gives a predetermined low response. Such models may relate the incidence of a quantal effect to dose, or they may consider the distribution of the "time to occurrence" of a condition and its relation to dose. In both cases, the results of extrapolation are strongly dependent on the choice of the model. For example, the Advisory Committee on Food Additives (FDA, 1970) noted that dose-response data may fit several models equally well in the 2% to 5% range, but the doses extrapolated to very low responses would differ very strikingly: the ratio of ED1 to ED0.000001 would be either 100,100000, or 1,000,000 for the probit, logistic, or one-hit curves, respectively.

Several extrapolation procedures have been proposed which will give an upper limit to the dose corresponding to a low response. In other words, the result of extrapolation will not be the best estimate of the unknown dose required to give the desired response but a dose that is most likely to be below the dose required to give this response. Two procedures based on this approach have received particular attention: one is based on the one-hit model, the other on the probit model.

The one-hit model assumes that an effect can be induced after a single susceptible target has been reached by a single biologically effective unit of dose. At low doses, this model is numerically equivalent to the linear dose-response model which is compatible with animal data for some carcinogens (Druckrey, 1967) and with some human data such as the incidence of lung cancer in relation to the number of cigarettes smoked per day. In their simplest form (*i.e.* when the true response at zero dose is assumed to be zero), the currently used extrapolation procedures based on this model operate as follows: (1) the upper 99% confidence limit (UCL) is estimated for the observed response at a dose d; (2) a desired limit is set for a low response (R) e.g. 1 in 1 000 000; and (3) the dose (de) that would produce a response which is, with a 99% probability, lower than R is calculated from the equation de = d*R/(UCL). Such procedures are more conservative than the procedures based on any other currently used dose-response model (probit, logit, or extreme-value models). In addition, the one-hit model seems to have a reasonable biological basis for carcinogenesis at low doses.

Researchers proposed the use of probits and a log-normal distribution to describe the variability of the sensitivities (tolerances) of individuals in a population. The probit model gives a dose-response curve that is concave at low-dose levels, and is less conservative than the linear model based on the one-hit hypothesis. The Mantel-Bryan procedure involves (1) the choice of a desired limit of response (R) (e.g. 1 in 1,000,000); (2) the estimation of the upper 99% confidence limit (UCL) for the observed response at dose d, and (3) imposing a probit-log dose straight line through UCL, with a slope (ß) equal to 1 (i.e. one probit per 10-fold dose-range). The choice of the slope (ß) is critical in this procedure. ß = 1 has been proposed because a slope greater than 1 is usually (but not always) observed in carcinogenesis experiments. The Mantel & Bryan procedure has been modified to take into account response levels in control groups.

A second category of models is based on the observation that the median "time to occurrence" (latent period) of an effect such as cancer may increase as the dose decreases but not proportionally. A thousand-fold change in the dose usually causes an approximately ten-fold change in the median time to tumour appearance and, with decreasing dose levels, a dose may be reached which would predict tumour occurrence beyond the life expectancy of the exposed individuals. This would still be consistent with the hypothesis that molecular changes in the cells, occurring in proportion to the concentration of carcinogens, are the initiating event. One model considers a log-normal distribution of the time-to-occurrence with median time depending on dose but with standard deviation independent of the dose. The dose d is related to the median time (t) by d= c/tn, where n is assumed to be greater than one, and c is a constant. Researchers compared this model with another model in which the time to occurrence is considered to have a Weibull distribution. They found that the Weibull distribution agreed with experimental data better than the log-normal, but this no doubt depends on the type of cancer involved. The low-dose extrapolation using these models would also be strongly dependent on the choice of the model. Dose-"time-to-occurrence"-response relationships in cancer risk assessment have also been considered by Janyseva & Antomonov.

The application of all these procedures presents practical difficulties. Low-dose extrapolation is thus a very difficult problem that cannot be solved by statistical methods alone. Great caution should be exercised in using the existing methods and their inherent limitations should always be kept in mind. Good experimental data, combined with human data if available, and an understanding of the mechanisms of toxic action are essential if the task of low-dose extrapolation is to be accomplished satisfactorily.

Other Methods of Extrapolation

A method for extrapolation from one species to another based on an established relationship between the indices of toxicity and body weight for different animal species has also been suggested. In mammals, the weights of internal organs, and many physiological variables (pulse and respiration rates, consumption of oxygen, food, and water, liver microsomal enzyme activity) show a log-log linear relationship with body size of the animals (allometric ratios). This regularity appears to be valid for more than 100 different variables including the period of gestation, litter size, erythrocyte life span, and latent period of tumour development but there are also other variables to which it does not apply. This "body weight rule" may be expressed as follows: the logarithms of biological variables of mammals show a linear regression to the logarithms of the body weight.

Researchers showed that values for the lethal dose for dogs of several chemicals obtained from regression analysis of toxicity in four other species of small mammals compared well with predictions made from direct extrapolation from albino rats or from the most sensitive animal species, or from the relationship of body surface area.

Human Data

Ethical Considerations

In research involving human subjects, a number of elements, such as the assessment of risk, potential benefit, and quality of consent, have to be evaluated to ascertain whether ethical considerations are satisfied. The essential provisions for protecting human subjects in experimentation and research have been expounded by many international and national organizations. Key factors include the right to informed consent and freedom from coercion. The international instruments in dealing with this matter are the Declaration of Helsinki and Article 7 of the International Covenant on Civil and Political Rights, adopted by the United Nations General Assembly, December 1966. Article 7 provides that "no-one shall be subjected without his free consent to medical or scientific experimentation". Some countries possess specific codes of ethics relating to human experimentation, and special problems of experimentation that involve the use of fetuses, children, the mentally ill, and prisoners require special consideration.

It is essential that human experimentation should only be undertaken when there is adequate evidence from animal and other studies that both the chemical and the circumstances of administration are safe. Every experiment with human volunteers should be subject to prior review and approval by a local ethical committee in order to ensure that the intended study complies with the ethical principles embodied in the Declaration of Helsinki and with other requirements of national and local bodies.

Ideal conditions of truly informed consent may not always be achieved in practice, consequently the burden of responsibility rests mainly with the investigator and, to a lesser extent, with the peer review body. Because of these difficulties, the guidelines and procedures for the protection of human subjects should be constantly reviewed and updated.

In any case, collection of data from human subjects must be accomplished with due respect for human rights and dignity. The use of ethics committees with broad representation to review and approve all such experimentation is recommended to protect the rights of human subjects and to ensure responsible investigation.

Need for Human Investigations

Although there is general repugnance at the idea of using human subjects to assess the safety of environmental chemicals, the question is not whether or not human subjects should be used in toxicity experiments but rather whether such chemicals, deemed from animal toxicity studies to be relatively safe, should be released first to controlled, carefully monitored groups of human subjects, instead of being released indiscriminately to large populations with no monitoring and with little or no opportunity to observe adverse effects.

The prediction and prevention of possible toxic hazards that may arise from the introduction of chemicals into the environment can be made more valid if data from studies of the chemical in human subjects are available. Three particular aspects of human toxicology have need of such information, namely: (a) the selection, through comparative consideration of metabolism, of the most appropriate animal species for studies to predict the human response; (b) investigation of a specific, reversible effect of the compound in the most sensitive animal species, to determine whether there is a correlation with a similar effect in man; and (c) study of effects specific to man.

Certain types of information about the effects of chemicals can only be obtained by direct observations on man. Often, carefully controlled experiments can provide significant information at doses well below those anticipated to be "safe"; measurement of subtle changes of reaction time, behavioural functions, and sensory responses may be examples. In other cases, useful information may be obtained by careful studies on human cells or tissue maintained by culture techniques.

Human toxicological data include both the data obtained from epidemiological surveys of populations exposed to a toxic chemical under normal conditions of use, in cases of acute accidental poisoning and in occupational exposure, and the data from experiments in volunteers. Although an experiment is defined as observations under controlled conditions of exposure, there is, at times, only a grey area that distinguishes an experiment with human subjects from observations on human subjects under natural conditions. For example, some segments of human populations are at higher risk and should be particularly closely monitored, e.g., those exposed to chemicals at work or those receiving continuous treatment with medicines. The periodic clinical evaluation of workers is normally the responsibility of the employer and careful records of these examinations coupled with measurement of exposure conditions often exist. If accidental excessive exposure of an individual or a population should occur, it is both ethical and pertinent to learn as much as possible, recognizing always the right of the patient. Because of the wide individual variation in the toxicity of chemicals to man, the final evaluation should be based on information obtained from as widely varied a human population as is compatible with the various ethical principles involved.

The Use of Toxicological Data in Establishing Environmental Health Standards

Environmental Health Standards

The aim of environmental health standards is to protect individuals, human populations, and their progeny from the adverse effects of hazardous environmental factors, including chemicals. A sound principle of health protection is to keep all exposures as low as reasonably achievable, subject to the condition that the appropriate exposure limits, defined by the standard, are not exceeded.

Environmental health standards for chemicals may be formulated either in terms of concentrations in environmental components (e.g., air, water, food, consumer products) or in terms of amounts of substances that may be taken into the body. These concentrations and amounts should be sufficiently low that the threshold dose (if it exists and can be determined) will not be reached, or that the population of concern will not be subject to "unacceptable" risk, even following life-time or working life-time exposure. In some cases, as for irritant air pollutants, the distribution of exposure concentrations in time should also be considered. Standards may also prescribe the quantity of a substance to be used at any time and the manner of its use.

Social, cultural, and economic considerations should be taken into account in setting standards, but never to the detriment of health protection which should be of primary concern.

It is obvious that a standard setting process will necessarily involve many considerations besides toxicology. This process is often very different in different countries and different types of society. In general, however, it involves appraisal of toxicological data, particularly of dose-response relationships, including the effects on non-human targets (plants, animals, materials); social and economic analysis, policy analysis and review of experience elsewhere, leading eventually to an administrative or policy decision concerning the standard. Other relevant questions include the technological feasibility of achieving a standard, the cost and benefit of implementing it, means of enforcement, other public health priorities etc. Many of these topics are outside the scope of the present monograph.

Assessment of Health Risk and Evaluation of Benefits

Assessment of health risk from a given exposure to an environmental factor is an essential step in any procedure for setting environmental health standards. Assessment of health risk involves more than routine application of "safety factors", or low dose extrapolation which provides estimates of response that are, strictly speaking, applicable only to the conditions of the experiment. Questions such as the incidence of effects in various age groups and the degree of life shortening in affected individuals are all relevant to standard setting. For this reason the study of "time-to-occurrence" models in the extrapolation of data should be encouraged. In addition, the seriousness of adverse effects will have to be evaluated from the public health and social viewpoint. Attention should also be paid to the heterogeneity of human populations, and, at present, it is not clear how the existence of susceptible groups, and the influence of nutrition and pre-existing disease in human populations should be taken into account. The existing methods of extrapolation from animal data to man deal with exposures to single substances whereas the actual human environment contains a large number of hazardous chemicals and other factors that can interact and considerably modify the effects, for example, in cancer induction. From this viewpoint, the importance of epidemiology and systematic surveillance of high risk groups cannot be overemphasized.

The acceptability of a given risk should also be considered in standard setting. This, as well as the judgment on safety (which involves decision on the acceptability of risk) exceeds the expertise of toxicologists. This is a domain where society at large has a role to play. Political decisions are also required on various social, economic, and ecological concerns. The same applies to the evaluation of benefits. As pointed out by a WHO Expert Committee "the expertise needed for the evaluation of risk is different from that needed for benefit evaluation. On the risk side, concern is focused on adverse health effects on man, damage to the environment, and misuse of natural resources. On the benefit side, the emphasis is on value to the consumer and the country". The interaction of all these factors is often described by the term "risk and benefit analysis", which is only partly within the area of toxicological expertise. The final judgment as to whether the benefit does or does not justify a risk is for society to make.

An example of Toxicological Information used in Standard Setting

Although standard setting procedures will differ from country to country, and the requirements for toxicological information will vary to a considerable extent, it may be useful, nevertheless, to describe, as an example, the procedure used in the USSR for setting standards for chemical pollutants in surface waters.

Information is first obtained on the likely concentrations of the chemical in industrial waste waters and the physical and chemical properties of the chemical. The stability of the substance under environmental conditions is then evaluated by standard analytical methods and the influence of the chemical on the self-purification processes of natural waters is studied.

The toxicological investigations required include LD50 studies for mice, rats, guineapigs and rabbits and subacute experiments lasting 1-2 months, to provide data on functional disturbances of organs and systems and on any cumulative properties of the chemical. These tests are followed by chronic toxicity experiments lasting 6-8 months. Study of specific effects of chemical water pollutants (e.g., mutagenicity, teratogenicity, and effects on reproductive function) is also a necessary component of toxicological investigations.

The subthreshold (maximum ineffective) concentration determined by chronic experiments is then compared with the threshold concentrations established for the other two indices of water quality (i.e. effects on the self-purification of water and its organoleptic properties), and the smallest concentration is assumed to be the "hygienic" standard.

The total number of hygienic standards for hazardous substances in water, developed in the USSR, has reached 500; of these, about 60% have been established according to organoleptic criteria and 30% according to the toxicological hazard index. These standards have been incorporated in the water legislation of the country and serve as the basis for practical control measures in protecting water bodies from chemical pollution.

Limitations of Safety Evaluation

Experimental toxicology is a highly complex, multidisciplinary science. The extrapolation of animal data to man requires well-informed contributions from several scientific disciplines. Absolute proof of safety for man of a chemical substance cannot be obtained from the results of toxicological tests. However, toxicological tests do provide guidance on the relative toxicity of a compound and help in identifying likely modes of action in man.

Acute toxicity studies in animals are of value in predicting potential toxic effects of a chemical in human beings exposed to near fatal doses. From the results of such studies, the nature of acute responses in man may be anticipated with a view to initiating life-supporting measures or first-aid or therapeutic procedures.

Short-term and subacute studies are particularly valuable in determining the more subtle toxic effects of a chemical, whether or not it has potential for cumulative toxicity and whether or not the toxic effects are reversible upon cessation of exposure. These tests are of value in estimating the potential hazard to man following exposure of intermediate duration, usually 2-7 years.

Of greatest concern to toxicologists, regulatory officials, and the general public are the possible chronic toxic effects of chemicals. Chronic toxicity tests assist in establishing the degree of risk to man that may be expected from low-level long-term exposure to a chemical substance. Chemicals that tend to persist or concentrate in the biosphere and as a result have the potential to affect large segments of the population are of particular concern.

In extrapolating animal data to man, several factors must be taken into consideration. These include the "no-effect level" derived from animal experiments, the nature of the dose-response curve and the nature of the toxic effects produced. The known or anticipated level of exposure in man and the potential number of exposed individuals must also be considered. It is worth pointing out that the so-called "no-effect level" is a statistically derived value usually estimated within a 95% confidence interval and that a 5% probability exists that the value is in error. It has been noted, for example, that if a toxic effect occurs in only 1% of the test animals, the effect will be entirely missed 37% of the time if only 100 animals are used in each test. In addition, if the same effect occurs spontaneously in control animals, the chances of detecting that response in treated animals becomes even more remote.

Predictions of toxicity from laboratory animal studies are dependent on the relevance of these studies to man, to wild-life, and to environmental ecosystems. They are also dependent on the genetics, nutrition, general health, and environmental circumstances of the individuals exposed.

There may be a hereditary disposition in man to an increased susceptibility to toxic chemicals, such as an increased tendency to malignant tumours. Similarly, persons under stress or treatment with immunosuppressive drugs may also be at greater risk to chemical toxicity and chemical carcinogenesis. These individuals will constitute abnormal populations for which the degree of risk may not be predictable from animal studies or from human studies carried out on healthy subjects, but these abnormal populations may be of sufficient magnitude to merit special consideration. Furthermore, genetic variations in laboratory animals are paralleled by variations in the toxic response to chemicals, and this puts additional limitations on predictions of possible human toxicity from such animal data.

Similarly, the nutritional status of individuals may also result in wide variation in susceptibility to toxic chemicals because malnutrition may lead to reduction of natural protection afforded by detoxication mechanisms.

Safety evaluation of chemicals is too frequently empirical and there is often a tendency to mistake quantity of data for quality. Regulatory agencies and toxicologists must be flexible and keep abreast of new experimental techniques and methods and of fundamental developments in the understanding of the mechanisms of toxicity. The application of new methods could be more relevant and informative than the routine use of old traditional ones, but these new methods should not necessarily be expected to replace traditional procedures, nor should they be applied routinely until adequately evaluated for significance and reliability.

Whether new or old procedures are employed, it is very important that the specific conditions under which the experiments are conducted should be accessible to other scientists, so that the results from different laboratories may be compared. Where it is not possible to set forth such details in publications of toxicological investigations, a central listing of detailed experimental procedures and conditions would be desirable.

ENTOMOTOXICOLOGY

In forensic entomology, entomotoxicology is the analysis of toxins in arthropods (mainly flies and beetles) that feed on carrion. Using arthropods in a corpse or at a crime scene, investigators can determine whether toxins were present in a body at the time of death. This technique is a major advance in forensics; previously, such determinations were impossible in the case of severely decomposed bodies devoid of intoxicated tissue and bodily fluids. Ongoing research into the effects of toxins on arthropod development has also allowed better estimations of postmortem intervals.

Effects of toxins on arthropods

Drugs can have a variety of effects on development rates of arthropods. Morphine, heroin, cocaine, and methamphetamine are commonly involved in cases where forensic entomology is used. The stages of growth for insects provides a basis for determining a cause in altered cycles in a specific species. An altered stage in development can often indicate toxins in the carrion on which the insects are feeding. Beetles (Order: Coleoptera) and beetle feces are often used in entomotoxicology, but the presence of toxins is often the result of the beetles' feeding on fly larvae that have been feeding on the carrion containing toxic substances. Flies (Order: Diptera) are the most commonly used insect in entomotoxicology.

Through the study of Sarcophaga (Curranea) tibialis larvae, barbiturates were found to increase the length of the larval stage of the fly, which will ultimately cause an increase in the time it takes to reach the stage of pupation. Morphine and heroin were both believed to slow down the rate of fly development. However, closer examination of the effects of heroin on fly development has shown that it actually speeds up larval growth and then decreases the development rate of the pupal stage. This actually increases the overall timing of development from egg to adult. Research of Lucilia sericata (Diptera: Calliphoridae), reared on various concentrations of morphine injected meat, found higher concentrations of morphine in shed pupal casings than in adults. Cocaine and methamphetamine also accelerate the rate of fly development.

Some effects depend on the concentration of the toxin while others simply depend on its presence. For example, cocaine (at the lethal dose) causes larvae to "develop more rapidly 36 (to 76) hours after hatching". The amount of growth depends on the concentration of cocaine in the area being fed upon. The amount of methamphetamine, on the other hand, affects the rate of pupal development. A lethal dose of methamphetamine increases larval development through approximately the first two days and afterwards the rate drops if exposure remains at the median lethal dosage. The presence of methamphetamine was also found to cause a decrease in the maximum length of the larvae.

Along with changes in development rates, extended periods of insect feeding refrain and variation in the size of the insect during any stage of development, can also indicate the presence of toxic substances in the insect's food source.

Examples of use

Since J.C. Beyer and his partners first demonstrated the ability of toxins to be recovered from maggots feeding on human remains in 1980, the use of entomotoxicology in investigations has made an emergence into the field of forensic entomology. An example of one such case involved the discovery of a 22-year-old female with a history of suicide attempts found 14 days after her death. Due to the body's advanced stage of decomposition, no organ or tissue samples were viable to screen for toxins. Through gas chromatography (GC) and thin-layer chromatography (TLC) analysis of Cochliomyia macellaria (Diptera: Calliphoridae) larvae found feeding on the woman's body, phenobarbital was detected and perceived to have been in the woman's system upon death.

Drug abuse detected

In France, Pascal Kintz and his colleagues were able to demonstrate the use of entomotoxicology to detect toxins that were not discovered during the analysis of body tissues and fluids of a body found roughly two months after death. A liquid chromatography analysis on organ tissue and Calliphoridae larvae found at the scene revealed the existence of five prescription medications. Triazolam, however, was only detected in the analysis of maggots and not in organ tissue samples. Comparative research showed increased sensitivity of toxicological analysis of Diptera samples over decomposed body tissues. A similar case involved the discovery of the remains of a 29-year-old known to abuse drugs, last seen alive five months prior. Through the use of GC and GC-MS techniques, Nolte and his partners discovered the presence of cocaine in decomposed muscle tissue and in maggots found on the body. However, due to the severity of decomposition of the muscle tissue, more suitable drug samples (devoid of decomposition byproducts) were reared from the maggots.

Aid determination of origin

Pekka Nuorteva presented the case of a young woman found severely decomposed in Ingå, Finland. Diptera larvae recovered from the body were reared to adulthood and found to contain low levels of mercury, indicating that the woman came from an area of comparatively low mercury pollution. This assumption was proven correct once the woman was identified and found to have been a student in Turku, Finland. This case demonstrated the ability of toxicological analysis to help determine origin. This case applied Nuorteva's research involving mercury and its effect on maggots. Through experimentation, it was determined that maggots (fed on fish containing mercury) possessed levels of mercury in their tissue of even greater concentration than in the tissue of the fish. Nuorteva also discovered that the presence of mercury in the maggots systems hindered their ability to enter into the pupal stage.

TOXIN CONFOUNDING OF POSTMORTEM INTERVAL ESTIMATE

Through the analysis of specific cases, it was revealed that toxins present in a person's body upon death can confound postmortem interval estimations. An example of such a case, reported by Gunatilake and Goff, concerned the discovery of a 58-year-old male with a history of attempted suicides found dead in a crawl space in Honolulu, Hawaii last seen eight days prior. Two species of Diptera (Calliphoridae), Chrysomya megacephala and Chrysomya rufifacies, found on the corpse and tissue samples from the body revealed malathion. Investigators found it abnormal that, given the conditions, there were only two fly species found on the body and that these species revealed a postmortem interval of five days. Thus it was determined that the presence of the organophosphate malathion in the man's system delayed oviposition for a few days.

Paul Catts analyzed a case in Spokane, Washington where maggots rendered differing postmortem estimations. A 20-year-old female victim was found stabbed to death and laying in an open environment surrounded by trees. Most of the oldest maggots found on the body were approximately 6–7 mm long which suggested that they were roughly seven days old. There was, however, a very strange exception which was the retrieval of a 17.7 mm maggot which suggested an age of 3 weeks. After ruling out the possibility that the maggot had traveled onto the corpse from carrion nearby, it was assumed that there was no conceivable way a 3-week-old maggot could have been present on the corpse. Later investigations revealed that the woman had snorted cocaine shortly before her death and that the 17.7 mm maggot must have fed in the woman's nasal cavity. Research revealed that maggot development can be sped up by the ingestion of cocaine.

Use of shed casings and insect faeces

Not only are tissues from maggots used to detect toxins, shed casings and insect faeces have also been used to detect and identify toxins present in corpses upon death. An instance of this finding was demonstrated by Edward McDonough, a medical examiner in Connecticut. A mummified corpse of a middle-aged woman was found inside of her home. Prescription medicine bottles were found with labels identifying the following drugs: ampicillin, Ceclor, doxycycline, erythromycin, Elavil, Lomotil, pentazocine, and Tylenol 3. McDonough performed toxicological analyses on stomach contents and dried sections of brain and found lethal levels of amitriptyline and nortriptyline. Insect feces, shed pupal cases of Megaselia scalaris (Diptera: Phoridae), and shed larval skins of Dermestes maculates (Coleoptera: Dermestidae) were gathered from the corpse at the scene. McDonough sent these to an FBI lab which broke down the complex structures of the samples using strong acids and bases and freed the toxins for analysis. The cast pupal cases and larval skins were also found to contain amitriptyline and nortriptyline. Larger concentrations were discovered in the pupal cases because phorid flies prefer to feed on softer tissues. The hide beetle larval skins revealed lower concentrations of the drugs because these beetles prefer to feed on dry, mummified bodies. The use of pupal cases and larval skins allows investigators to detect toxins in a body years after death.

Limitations

Further research should be conducted in order to fill the gaps in entomotoxicology. Such areas as bioaccumulation, insect metabolism of drugs, and quantitative analyses of insect evidence have only begun to be researched. Because it is a relatively new branch of forensic entomology, entomotoxicology has its limitations. According to Pounder's research, there is no correlation between the drug concentration in tissue and the larvae feeding on that tissue. Entomological specimens make for excellent qualitative toxicological specimens. There is, however, a lack of research in the way of developing an assessment that can quantify the concentration of a drug in tissue using entomological evidence. One reason for this is that a drug can only be detected in larvae when the rate of absorption exceeds the rate of elimination. demonstrated this theory using Calliphora vicina larvae reared on human skeletal muscle obtained from cases of co-proxamol and amitriptyline overdose. Samples of pupae and third instar larvae no longer contained concentrations of the drugs, suggesting that drugs do not bioaccumulate over the entire life-cycle of larvae. This leads entomologists to theorize that toxins are eliminated from the larvae's system over time if they are not receiving a constant supply of the toxin.

Automatism (toxicology)

Automatism, in toxicology, refers to a tendency to take a drug over and over again, forgetting each time that one has already taken the dose. This can lead to a cumulative overdose. A particular example is barbiturates which were once commonly used as hypnotic (sleep inducing) drugs. Among the current hypnotics, benzodiazepines, especially midazolam might show marked automatism, possibly through their intrinsic anterograde amnesia effect. Barbiturates are known to induce hyperalgesia, i.e. aggravation of pain and for sleeplessness due to pain, if barbiturates are used, more pain and more disorientation would follow leading to drug automation and finally a "pseudo"suicide. Such reports dominated the medical literature of 1960s and 1970s; a reason replacing the barbiturates with benzodiazepines when they became available.

Aquatic toxicology

Aquatic toxicology is the study of the effects of manufactured chemicals and other anthropogenic and natural materials and activities on aquatic organisms at various levels of organization, from subcellular through individual organisms to communities and ecosystems. Aquatic toxicology is a multidisciplinary field which integrates toxicology, aquatic ecology and aquatic chemistry.

This field of study includes freshwater, marine water and sediment environments. Common tests include standardized acute and chronic toxicity tests lasting 24–96 hours (acute test) to 7 days or more (chronic tests). These tests measure endpoints such as survival, growth, reproduction, that are measured at each concentration in a gradient, along with a control test. Typically using selected organisms with ecologically relevant sensitivity to toxicants and a well-established literature background. These organisms can be easily acquired or cultured in lab and are easy to handle.

While basic research in toxicology began in multiple countries in the 1800s, it was not until around the 1930s that the use of acute toxicity testing, especially on fish, was established. Due to the popularity of organochlorine pesticide DDT [l,l,l-trichloro-2,2-bis(p-chlorophenyl)ethane] and its linkage to causing fish death, the field of aquatic toxicology grew. At first, studies focused mainly on oysters and mussels, as they could not move away from the toxic environment. Over the next two decades, the effects of chemicals and wastes on non-human species became more of a public issue and the era of the pickle-jar bioassays began as efforts increased to standardize toxicity testing techniques.

In the United States, the passage of the Federal Water Pollution Control Act of 1947 marked the first comprehensive legislation for the control of water pollution and was followed by the Federal Water Pollution Control Act in 1956. In 1962, public and governmental interests were renewed, in large part due to the publication of Rachel Carson's Silent Spring, and three years later the Water Quality Act of 1965 was passed, which directed states to develop water quality standards. Public awareness, as well as scientific and governmental concern, continued to grow throughout the 1970s and by the end of the decade research had expanded to include hazard evaluation and risk analysis. In the subsequent decades, aquatic toxicology has continued to expand and internationalize so that there is now a strong application of toxicity testing for environmental protection.

Currently, aquatic toxicology is continuing to evolve as risk assessment is becoming more practiced in the field. The field is gaining popularity as it has begun to link the effects of pollutants on marine animals to humans who eat fish and other marine life.

Aquatic toxicity tests

Aquatic toxicology tests (assays): toxicity tests are used to provide qualitative and quantitative data on adverse (deleterious) effects on aquatic organisms from a toxicant. Toxicity tests can be used to assess the potential for damage to an aquatic environment and provide a database that can be used to assess the risk associated within a situation for a specific toxicant. Aquatic toxicology tests can be performed in the field or in the laboratory. Field experiments generally refer to multiple species exposure, but single species can be caged for a set duration, and laboratory experiments generally refer to single species exposure. A dose–response relationship is most commonly used with a sigmoidal curve to quantify the toxic effects at a selected end-point or criteria for effect (i.e. death or other adverse effect to the organism). Concentration is on the x-axis and percent inhibition or response is on the y-axis.

The criteria for effects, or endpoints tested for, can include lethal and sublethal effects (see Toxicological effects).

There are different types of toxicity tests that can be performed on various test species. Different species differ in their susceptibility to chemicals, most likely due to differences in accessibility, metabolic rate, excretion rate, genetic factors, dietary factors, age, sex, health and stress level of the organism. Common standard test species are the fathead minnow (Pimephales promelas), daphnids (Daphnia magna, D. pulex, D. pulicaria, Ceriodaphnia dubia), midge (Chironomus tentans, C. riparius), rainbow trout (Oncorhynchus mykiss), sheepshead minnow (Cyprinodon variegatu), zebra fish (Danio rerio), mysids (Mysidopsis), oyster (Crassotreas), scud (Hyalalla Azteca), grass shrimp (Palaemonetes pugio) and mussels (Mytilus galloprovincialis). As defined by ASTM, these species are routinely selected on the basis of availability, commercial, recreational, and ecological importance, past successful use, and regulatory use.

A variety of acceptable standardized test methods have been published. Some of the more widely accepted agencies to publish methods are: the American Public Health Association, US Environmental Protection Agency (EPA), ASTM International, International Organization for Standardization, Environment and Climate Change Canada, and Organisation for Economic Co-operation and Development. Standardized tests offer the ability to compare results between laboratories. There are many kinds of toxicity tests widely accepted in the scientific literature and regulatory agencies. The type of test used depends on many factors: Specific regulatory agency conducting the test, resources available, physical and chemical characteristics of the environment, type of toxicant, test species available, laboratory vs. field testing, end-point selection, and time and resources available to conduct the assays are some of the most common influencing factors on test design.

Exposure systems

Exposure systems are four general techniques the controls and test organisms are exposed to the dealing with treated and diluted water or the test solutions.

Static. A static test exposes the organism in still water. The toxicant is added to the water in order to obtain the correct concentrations to be tested. The control and test organisms are placed in the test solutions and the water is not changed for the entirety of the test.

Recirculation. A recirculation test exposes the organism to the toxicant in a similar manner as the static test, except that the test solutions are pumped through an apparatus (i.e. filter) to maintain water quality, but not reduce the concentration of the toxicant in the water. The water is circulated through the test chamber continuously, similar to an aerated fish tank. This type of test is expensive and it is unclear whether or not the filter or aerator has an effect on the toxicant.

Renewal. A renewal test also exposes the organism to the toxicant in a similar manner as the static test because it is in still water. However, in a renewal test the test solution is renewed periodically (constant intervals) by transferring the organism to a fresh test chamber with the same concentration of toxicant.

Flow-through. A flow-through test exposes the organism to the toxicant with a flow into the test chambers and then out of the test chambers. The once-through flow can either be intermittent or continuous. A stock solution of the correct concentrations of contaminant must be previously prepared. Metering pumps or diluters will control the flow and the volume of the test solution, and the proper proportions of water and contaminant will be mixed.

Types of tests

Acute tests are short-term exposure tests (hours or days) and generally use lethality as an endpoint. In acute exposures, organisms come into contact with higher doses of the toxicant in a single event or in multiple events over a short period of time and usually produce immediate effects, depending on absorption time of the toxicant. These tests are generally conducted on organisms during a specific time period of the organism's life cycle, and are considered partial life cycle tests. Acute tests are not valid if mortality in the control sample is greater than 10%. However, this control acceptability criterion is dependent upon the species and the duration of the test. Results are reported in EC50, or concentration that will affect fifty percent of the sample size.

Chronic tests are long-term tests (weeks, months years), relative to the test organism's life span (>10% of life span), and generally use sub-lethal endpoints. In chronic exposures, organisms come into contact with low, continuous doses of a toxicant. Chronic exposures may induce effects to acute exposure, but can also result in effects that develop slowly. Chronic tests are generally considered full life cycle tests and cover an entire generation time or reproductive life cycle ("egg to egg"). Chronic tests are not considered valid if mortality in the control sample is greater than 20%. These results have generally been reported in NOECs (No observed effects level) and LOECs (Lowest observed effects level). However, NOECs and LOECs are becoming less common as endpoints are dependent on the concentration series chosen for the test. These reports are starting to become a topic of debate in the field because of the way it may alter the results of the tests. For example, if the concentration rate of the NOEC is 100, 50, 25, 11.25, 6.25 and the toxicology is reported at 2%, the NOEC would report the concentration as 6.25.

Early life stage tests are considered as subchronic exposures that are less than a complete reproductive life cycle and include exposure during early, sensitive life stages of an organism. These exposures are also called critical life stage, embryo-larval, or egg-fry tests. Early life stage tests are not considered valid if mortality in the control sample is greater than 30%.

Short-term sublethal tests are used to evaluate the toxicity of effluents to aquatic organisms. These methods are developed by the EPA, and only focus on the most sensitive life stages. Endpoints for these test include changes in growth, reproduction and survival. NOECs, LOECs and EC50s are reported in these tests.

Bioaccumulation tests are toxicity tests that can be used for hydrophobic chemicals that may accumulated in the fatty tissue of aquatic organisms. Toxicants with low solubilities in water generally can be stored in the fatty tissue due to the high lipid content in this tissue. The storage of these toxicants within the organism may lead to cumulative toxicity. Bioaccumulation tests use bioconcentration factors (BCF) to predict concentrations of hydrophobic contaminants in organisms. The BCF is the ratio of the average concentration of test chemical accumulated in the tissue of the test organism (under steady state conditions) to the average measured concentration in the water.

Freshwater tests and saltwater tests have different standard methods, especially as set by the regulatory agencies. However, these tests generally include a control (negative and/or positive), a geometric dilution series or other appropriate logarithmic dilution series, test chambers and equal numbers of replicates, and a test organism. Exact exposure time and test duration will depend on type of test (acute vs. chronic) and organism type. Temperature, water quality parameters and light will depend on regulator requirements and organism type.

In the US, many wastewater dischargers (e.g., factories, power plants, refineries, mines, municipal sewage treatment plants) are required to conduct periodic whole effluent toxicity (WET) tests under the National Pollutant Discharge Elimination System (NPDES) permit program, pursuant to the Clean Water Act. For facilities discharging to freshwater, effluent is used to perform static-acute multi-concentration toxicity tests with Ceriodaphnia dubia (water flea) and Pimephales promelas (fathead minnow), among other species. The test organisms are exposed for 48 hours under static conditions with five concentrations of the effluent. The major deviation in the short-term chronic effluent toxicity tests and the acute effluent toxicity tests is that the short-term chronic test lasts for seven days and the acute test lasts for 48 hours. For discharges to marine and estuarine waters, the test species used are sheepshead minnow (Cyprinodon variegatus), inland silverside (Menidia beryllina), Americamysis bahia, and purple sea urchin (Strongylocentrotus purpuratus).

Sediment tests

At some point most chemicals originating from both anthropogenic and natural sources accumulate in sediment. For this reason, sediment toxicity can play a major role in the adverse biological effects seen in aquatic organisms, especially those inhabiting benthic habitats. A recommended approach for sediment testing is to apply the sediment quality triad (SQT) which involves simultaneously examining sediment chemistry, toxicity, field alterations, bioaccumulation, and bioavailability assessments that can be used in a laboratory or in the field. Due to the expansion of SQTs, it is now more commonly referred to as "Sediment Assessment Framework." Collection, handling, and storage of sediment can have an effect on bioavailability and for this reason standard methods have been developed to suit this purpose.

Toxicological effects

Toxicity can be broken down into two broad categories of direct and indirect toxicity. Direct toxicity results from a toxicant acting at the site of action in or on the organism. Indirect toxicity occurs with a change in the physical, chemical, or biological environment.

Lethality is most common effect used in toxicology and used as an endpoint for acute toxicity tests. While conducting chronic toxicity tests sublethal effects are endpoints that are looked at. These endpoints include behavioral, physiological, biochemical, histological changes.

There are a number of effects that occur when an organism is simultaneously exposed to two or more toxicants. These effects include additive effects, synergistic effects, potentiation effects, and antagonistic effects. An additive effect occurs when combined effect is equal to a combination or sum of the individual effects. A synergistic effect occurs when the combination of effects is much greater than the two individual effects added together. Potentiation is an effect that occurs when an individual chemical has no effect is added to a toxicant and the combination has a greater effect than just the toxicant alone. Finally, an antagonistic effect occurs when a combination of chemicals has less of an effect than the sum of their individual effects.

Important aquatic toxicology resources

ASTM International (formerly American Society for Testing and Materials). A consensus-based organization, representing 135 countries, that develops and delivers international voluntary standard methods for aquatic toxicity testing.

Standard Methods for the Examination of Water and Wastewater. A compilation of techniques for water analysis, jointly published by the American Public Health Association (APHA), the American Water Works Association (AWWA), and the Water Environment Federation.

"Ecotox." A database maintained by EPA that offers single chemical toxicity information for both aquatic and terrestrial purposes.

Society of Environmental Toxicology and Chemistry (SETAC). A nonprofit, worldwide society working to promote scientific research to further our understanding of environmental stressors, environmental education, and the use of science in environmental policy.

US EPA publishes guidance manuals outlining aquatic toxicity test procedures.

Organisation for Economic Co-operation and Development (OECD). A forum for governments to work together to promote policies for the betterment of people's social and economic well-being around the world. One way in which they accomplish this is through the development of aquatic toxicity test guidelines.

Environment and Climate Change Canada. Canada's lead federal agency for environmental protection.

Terminology

Median Lethal Concentration (LC50) – The chemical concentration that is expected to kill 50% of a group of organisms.

Median Effective Concentration (EC50) – The chemical concentration that is expected to have one or more specified effects in 50% of a group of organisms.

Critical Body Residue (CBR) – An approach that routinely examines whole-body chemical concentrations of an exposed organism that is associated with an adverse biological response.

Baseline toxicity – Refers to narcosis which is a depression in biological activity due to toxicants being present in the organism.

Biomagnification – The process by which the concentration of a chemical in the tissues of an organism increases as it passes through several levels in the food web.

Lowest Observed Effect Concentration (LOEC) – The lowest test concentration that has a statistically significant effect over a specified exposure time.

No Observed Effect Concentration (NOEC) – The highest test concentration for which no effect is observed relative to a control over a specified exposure time.

Maximum Acceptable Toxicant Concentration (MATC) – An estimated value that represents the highest "no-effect" concentration of a specific substance within the range including the NOEC and LOEC.

Application Factor (AF) – An empirically derived "safe" concentration of a chemical.

Biomonitoring – The consistent use of living organisms to analyze environmental changes over time.

Effluent – Liquid, industrial discharge that usually contain varying chemical toxicants.

Quantitative Structure-Activity Relationship (QSAR) – A method of modeling the relationship between biological activity and the structure of organic chemicals.

Mode of Action – A set of common behavioral or physiological signs that represent a type of adverse response.

Mechanism of Action – The detailed events that take place at the molecular level during an adverse biological response.

KOW – The octanol-water partition coefficient which represents the ratio of the concentration of octanol to the concentration of chemical in the water.

Bioconcentration Factor (BCF) – The ratio of the average chemical concentration in the tissues of the organism under steady-state conditions to the average chemical concentration measured in the water to which the organisms are exposed.

All terms were derived from Rand.

2

ENVIRONMENTAL TOXICOLOGY

Environmental toxicology is a multidisciplinary field of science concerned with the study of the harmful effects of various chemical, biological and physical agents on living organisms. Ecotoxicology is a subdiscipline of environmental toxicology concerned with studying the harmful effects of toxicants at the population and ecosystem levels. Rachel Carson is considered the mother of environmental toxicology, as she made it a distinct field within toxicology in 1962 with the publication of her book Silent Spring, which covered the effects of uncontrolled pesticide use. Carson's book was based extensively on a series of reports by Lucille Farrier Stickel on the ecological effects of the pesticide DDT.

Organisms can be exposed to various kinds of toxicants at any life cycle stage, some of which are more sensitive than others. Toxicity can also vary with the organism's placement within its food web. Bioaccumulation occurs when an organism stores toxicants in fatty tissues, which may eventually establish a trophic cascade and the biomagnification of specific toxicants. Biodegradation releases carbon dioxide and water as by-products into the environment. This process is typically limited in areas affected by environmental toxicants.

Harmful effects of such chemical and biological agents as toxicants from pollutants, insecticides, pesticides, and fertilizers can affect an organism and its community by reducing its species diversity and abundance. Such changes in population dynamics affect the ecosystem by reducing its productivity and stability. Although legislation implemented since the early 1970s had intended to minimize harmful effects of environmental toxicants upon all species, McCarty (2013) has warned that "longstanding limitations in the implementation of the simple conceptual model that is the basis of current aquatic toxicity testing protocols" may lead to an impending environmental toxicology "dark age".

GOVERNING POLICIES ON ENVIRONMENTAL TOXICITY

International Policies

To protect the environment, The National Environmental Policy Act (NEPA) was written. The main point that NEPA brings to light is that it "assures that all branches of government give proper consideration to the environment prior to undertaking any major federal actions that significantly affect the environment." This law was passed in 1970 and also founded the Council on Environmental Quality (CEQ). The importance of CEQ was that it helped further push policy areas. CEQ created helpful environmental programs that were very beneficial. Some include, the Federal Water Pollution Control Act (RCRA), Toxic Substance Control Act, Resources Conservation and Recovery Act (RCRA and the Safe). CEQ was essential in creating the foundation for most of the "current environmental legislation except for Superfund and asbestos control legislation." Some initial impacts of NEPA pertain to the interpretation within Courts. Not only did Courts interpret NEPA to expand over direct environmental impacts from any projects, specifically federal, but also indirect actions from federal projects.

Toxic Substance Control Act

TSCA, also known as the Toxic Substance Control Act, is a federal law that regulates industrial chemicals that have the potential to be harmful to humans and the environment. TSCA specifically targets "the manufacture, importation, storage, use, disposal, and degradation of chemicals in commercial use." The EPA allows the following to be done:"

1. Premanufacture testing of chemicals to determine health or environmental risk
2. Review of chemicals for significant risk prior to the start of commercial production
3. Restriction or prohibition on the production or disposal of certain chemicals
4. Import and export control of chemicals prior to their entering or leaving the USA."

The Clean Air Act

The Clean Air Act was aided by the signing of the 1990 amendments. These amendments protected reducing acid, the ozone layer, improving air quality and toxic pollutants. The Clean Air Act was actually revised and with, support from

President George H.W Bush, it was signed in. The biggest major threats that this act targets are: urban air pollutions, toxic air emissions, stratospheric ozone, acid rain etc. Apart from targeting these specific areas, it also established a national operating that "permits program to make the law more workable, and strengthened enforcement to help ensure better compliance with the Act."

Regulations and Enforcement Actions on PCBs

As mentioned above, though the United States did ban the use of PCBs, there is the possibility that they are present in products made before the PBC band in 1979. The Environmental Protection Agency released its ban on PCBs on April 19th, 1979.According to the US Environmental Protection Agency, "Although PCBs are no longer being produced in this country, we will now bring under control the vast majority of PCBs still in use," said EPA Administrator Douglas M. Castle. "This will help prevent further contamination of our air, water and food supplies from a toxic and very persistent man-made chemical."

PCBs has been tested on laboratory animals and have caused cancer and birth defects. PCB is suspected of having certain effects on liver and skin of humans. They are also suspected of causing cancer as well. EPA, "estimates that 150 million pounds of PCBs are dispersed throughout the environment, including air and water supplies; an additional 290 million pounds are located in landfills in this country." Again, even though they have been banned, there is still a large amount of PCBs are circulating within the environment and are possibly causing effects on the skin and liver of humans. There were some cases in which people or companies that disposed of PCBs incorrectly. Up until now, there have been four cases in which EPA had to take legal actions against people/companies for their methods of disposal. The two cases involving the companies, were fined $28,600 for improper disposal. It is unknown what fined was charged against the three people for, "illegally dumping PCBs along 210 miles of roadway in North Carolina."

Though PCBs were banned, there are some exceptions where they are being used. The area in which it has been completely prohibited is, "the manufacture, processing, distribution in commerce, and "non-enclosed" (open to the environment) uses of PCBs unless specifically authorized or exempted by EPA. "Totally enclosed" uses (contained, and therefore exposure to PCBs is unlikely) will be allowed to continue for the life of the equipment." In terms of electrical equipment containing PCBs is allowed under specific controlled conditions. Out tof the 750 million pounds of PCBs, electrical equipment represents 578 million pounds. Any new manufacture of PCB is actually prohibited though.

Sources of environmental toxicity

There are many sources of environmental toxicity that can lead to the presence of toxicants in our food, water and air. These sources include organic and inorganic pollutants, pesticides and biological agents, all of which can have harmful effects on living organisms. There can be so called point sources of pollution, for instance the drains from a specific factory but also non-point sources (diffuse sources) like the rubber from car tires that contain numerous chemicals and heavy metals that are spread in the environment.

PCBs

Polychlorinated biphenyls (PCBs) are organic pollutants that are still present in our environment today, despite being banned in many countries, including the United States and Canada. Due to the persistent nature of PCBs in aquatic ecosystems, many aquatic species contain high levels of this chemical. For example, wild salmon (Salmo salar) in the Baltic Sea have been shown to have significantly higher PCB levels than farmed salmon as the wild fish live in a heavily contaminated environment. PCBs pertains to a group of human-produced"organic chemicals known as Chlorinated hydrocarbons" The chemical and physical properties of a PCS determine the quantity and location chlorine and unlike other chemicals, they have no form of identification. The range of toxicity is not consistent and because PCBs have certain properties (chemical stability, non-flammability) they have been used in a colossal amount of commercial and industrial practices. Some of those include, "Electrical, heat transfer and hydraulic equipment, plasticizers in paints, plastics and rubber products and pigments, dyes and carbonless copy paper" to name a few.

Heavy metals

Heavy metals found in food sources, such as fish can also have harmful effects. These metals can include mercury, lead and cadmium. It has been shown that fish (i.e. rainbow trout) are exposed to higher cadmium levels and grow at a slower rate than fish exposed to lower levels or none. Moreover, cadmium can potentially alter the productivity and mating behaviours of these fish. Heavy metals can not only affect behaviors, but also the genetic makeup in aquatic organisms. In Canada, a study examined genetic diversity in wild yellow perch along various heavy metal concentration gradients in lakes polluted by mining operations.

Researchers wanted to determine as to what effect metal contamination had on evolutionary responses among populations of yellow perch. Along the gradient, genetic diversity over all loci was negatively correlated with liver cadmium contamination. Additionally, there was a negative correlation observed between copper contamination and genetic diversity. Some aquatic species have evolved heavy metal tolerances. In response to high heavy metal concentrations a Dipteran species, Chironomus riparius, of the midge family, Chironomidae, has evolved to become tolerant to Cadmium toxicity in aquatic environments. Altered life histories, increased Cd excretion, and sustained growth under Cd exposure is evidence that shows that Chironomus riparius exhibits genetically based heavy metal tolerance.

Metals Toxicity

The most known or common types of heavy metals include zinc, arsenic, copper, lead, nickel, chromium and cadmium. All of these types cause certain risks on human and environment health. Though certain amount of these metals can actually have an important role in, for example, maintaining certain biochemical and physiological, "functions in living organisms when in very low concentrations, however they become noxious when they exceed certain threshold concentrations."Heavy metal are a huge part of environmental pollutions and their toxicity "is a problem of increasing significance for ecological, evolutionary, nutritional and environmental reasons."

Arsenic

Arsenic, one of the most important heavy metals, causes health problems within ecological and humans. It is, "semimetallic property, is prominently toxic and carcinogenic, and is extensively available in the form of oxides or sulfides or as a salt of iron, sodium, calcium, copper, etc." Not only that, but it is also the most abundant element here on earth and its specific inorganic forms are very dangerous to living creatures (animals, plants, and humans) and the environment. Effects of Arsenic on humans is that they can cause cancer in the bladder, skin, lungs and liver. One of the major areas in which humans are exposed to arsenic is through contaminated water which is a problem in more than 30 countries in the world.

Humans tend to encounter arsenic by, "natural means, industrial source, or from unintended sources." Water can become contaminated by arsenical pesticides or just natural arsenical chemicals. There are some cases in which arsenic has been used in suicidal attempts and can result in acute poisoning. Arsenic, "is a protoplastic poison since it affects primarily the sulphydryl group of cells causing malfunctioning of cell respiration, cell enzymes and mitosis."

Lead

Another extremely toxic metal, lead can and has been known to cause, "extensive environmental contamination and health problems in many parts of the world." The physical appearance of lead is bright and silver colored metal. Some sources of lead pollution in the environment include Metal plating and fishing operations, soil waste, factory chimneys, smelting of ores, wastes from batter industries, fertilizers and pesticides and many more. Unlike, other metals such as copper, lead only plays a physiological aspect and no biological functions. In the US, "more than 100 to 200,000 tons of lead per year is being released from vehicle exhausts" and some can be brought in by plants, flow in water or fixation into the soil. Humans come in contact with lead through mining, fossil fuel burning. In burning, lead and its compounds are exposed into air, soil, and water. Lead can have different effects on the body and effects the central nervous system. Someone who has come in contact with lead can have either acute or chronic lead poisoning. Those who experience acute poisoning have symptoms such as, appetite, headache, hypertension, abdominal pain, renal dysfunction, fatigue, sleeplessness, arthritis, hallucinations and vertigo." Chronic exposure on the other hand, can cause more severe symptoms such as, "mental retardation, birth defects, psychosis, autism, allergies, dyslexia, weight loss, hyperactivity, paralysis, muscular weakness, brain damage, kidney damage and may even cause death."

Mercury

Equally as toxic as the previous heavy metals, Mercury, A shiny silver-white, can transform into a colorless and odorless gas when heated up. Mercury highly affects the marine environment and there have been many studies conducted on the effects on the water environment. The biggest sources of mercury pollution include, "agriculture, municipal wastewater discharges, mining, incineration, and discharges of industrial wastewater" all relatively connected to water. Mercury exists in three different forms and all three possess different levels of bioavailability and toxicity. The three forms include, organic compounds, metallic elements and inorganic salts. As stated above, they are present in water resources such as oceans, rivers and lakes. They are absorb by microorganism, and go through, "biomagnification causing significant disturbance to aquatic lives. " Mercury hurts marine life but can also be very hurtful towards humans' nervous system. Higher levels of mercury exposure can change many brain functions. It can, "lead to shyness, tremors, memory problems, irritability, and changes in vision or hearing."

Cadmium

According to, ATSDR ranking, cadmium is the 7th most toxic heavy metal. Cadmium is interesting in that once it is exposed to humans or animals in their environment, it will accumulate inside the body throughout the life of the human/animal. Though cadmium was used as replacement for tin in WWI and pigment in paint industries back in the day, now day it is seen mostly in rechargeable batteries, tobacco smoke and some alloys production. As stated by the Agency for Toxic Substance and Disease Registry, in " the US, more than 500,000 workers get exposed to toxic cadmium each year." It is also stated that the highest exposure to Cadmium can be seen in China and Japan. The effects of Cadmium on the kidney and bones is huge. It can cause bone mineralization which " is the process of laying down minerals on a matrix of the bone". This can happen through renal dysfunction or bone damage.

Chromium

The 7th most abundant element, Chromium, can occur naturally when one burns oil and coal and is release into the environment through sewage and fertilizers. Chromium usage can be seen in, "industries such as metallurgy, electroplating, production of paints and pigments, tanning, wood preservation, chemical production and pulp and paper production." Chromium toxicity affects the "biological processes in various plants such as maize, wheat, barley, cauliflower, citrullus and in vegetables. Chromium toxicity causes chlorosis and necrosis in plants."

Pesticides

Pesticides are a major source of environmental toxicity. These chemically synthesized agents have been known to persist in the environment long after their administration. The poor biodegradability of pesticides can result in bioaccumulation of chemicals in various organisms along with biomagnification within a food web. Pesticides can be categorized according to the pests they target. Insecticides are used to eliminate agricultural pests that attack various fruits and crops. Herbicides target herbal pests such as weeds and other unwanted plants that reduce crop production.

DDT

Dichlorodiphenyltrichloroethane (DDT) is an organochlorine insecticide that has been banned due to its adverse effects on both humans and wildlife. DDT's insecticidal properties were first discovered in 1939. Following this discovery, DDT was widely used by farmers in order to kill agricultural pests such as the potato beetle, coddling moth and corn earworm. In 1962, the harmful effects of the widespread and uncontrolled use of DDT were detailed by Rachel Carson in her book The Silent Spring. Such large quantities of DDT and its metabolite Dichlorodiphenyldichloroethylene (DDE) that were released into the environment were toxic to both animals and humans.

DDT is not easily biodegradable and thus the chemical accumulates in soil and sediment runoff. Water systems become polluted and marine life such as fish and shellfish accumulate DDT in their tissues. Furthermore, this effect is amplified when animals who consume the fish also consume the chemical, demonstrating biomagnification within the food web. The process of biomagnification has detrimental effects on various bird species because DDT and DDE accumulate in their tissues inducing egg-shell thinning. Rapid declines in bird populations have been seen in Europe and North America as a result.

Humans who consume animals or plants that are contaminated with DDT experience adverse health effects. Various studies have shown that DDT has damaging effects on the liver, nervous system and reproductive system of humans. By 1972, the United States Environmental Protection Agency (EPA) banned the use of DDT in the United States. Despite the regulation of this pesticide in North America, it is still used in certain areas of the world. Traces of this chemical have been found in noticeable amounts in a tributary of the Yangtze River in China, suggesting the pesticide is still in use in this region. Though DDT was banned in 1972, some of the pesticide (as well as other chemical) lingered in the environment. This lingering of toxic material led to the near extinction of peregrine falcon. There was high levels of DDT were found in many areas such as "eggs, fat and tissues of the bird." The government . worked with conservation groups in helping them breed out of the contaminated area. Finally, in 1999 the birds were taken off the U.S endangered species list.

Sulfuryl fluoride

Sulfuryl fluoride is an insecticide that is broken down into fluoride and sulfate when released into the environment. Fluoride has been known to negatively affect aquatic wildlife. Elevated levels of fluoride have been proven to impair the feeding efficiency and growth of the common carp (Cyprinus carpio). Exposure to fluoride alters ion balance, total protein and lipid levels within these fish, which changes their body composition and disrupts various biochemical processes.

Cyanobacteria and cyanotoxins

Cyanobacteria, or blue-green algae, are photosynthetic bacteria. They grow in many types of water. Their rapid growth ("bloom") is related to high water temperature as well as eutrophication (resulting from enrichment with minerals and nutrients often due to runoff from the land that induces excessive growth of these algae). Many genera of cyanobacteria produce several toxins. Cyanotoxins can be dermatotoxic, neurotoxic, and hepatotoxic, though death related to their exposure is rare. Cyanotoxins and their non-toxic components can cause allergic reactions, but this is poorly understood. Despite their known toxicities, developing a specific biomarker of exposure has been difficult because of the complex mechanism of action these toxins possess.

OCCURRENCE OF CYANOTOXINS IN DRINKING WATER

The occurrence of this toxin in drinking water depends on a couple of factors. One, is the drinking water's level in raw source water and secondly, it depends on the effectiveness of removing these toxins from water when drinking water is actually being produced. Due to being no data on the absence/presence of these toxins in drinking water, it is very hard to actually monitor the amounts that are present in finished water. This is a result of the U.S not having an state or federal programs in place that actually monitor the presence of this toxins in drinking water treatment plants.

Effects of Toxins on Humans

Though data on the effects of these two toxins are limited, from what is available it suggests the toxins attack the liver and kidney. There was an hepatoenteritis-like outbreak in Palm Island, Australia (1979) due to the consumption of water that contained, "C. raciborskii, a cyanobacteria that can produce cylindrospermopsin." Most cases (typically involving children) have required they be taken to a hospital. The results of hospitilation include: Vomiting, kidney damage (due to lose of water, protein and electrolytes) fever, bloody diarrhea, and headaches.

Risk assessment

Environmental hazards posed by a substance are assessed according to three factors:

- P Persistence in the environment (bioaccumulation)
- B Affinity for fat (substances with high affinity for fat accumulate in the body)
- T Toxic to fish or other animals

In deciding whether a substance poses a threat to the environment regulatory organisations need to consider the overall level of exposure. Concerns over a substance are usually identified through epidemiology and wildlife studies. Once the risks are understood further tests such in vitro studies on cells or reproductive studies on animals are considered.

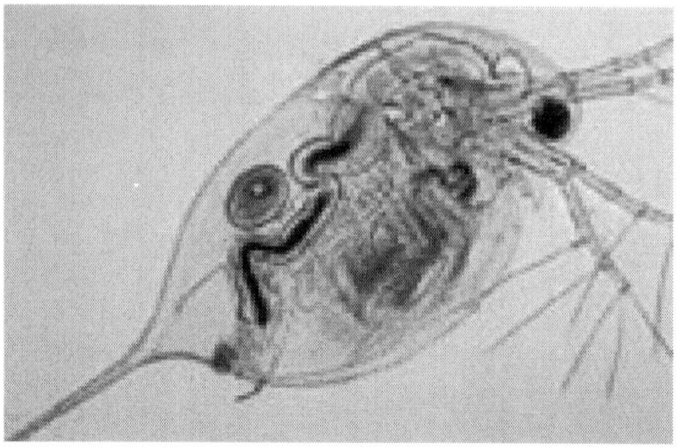

Daphnia species are commonly used in environmental toxicity tests

A risk-benefit analysis is used to protect the environment from harmful substances while allowing material and chemical technologies to be used effectively and developed. Animal tests are used to understand the potential damage that may be caused by a particular substance, to show how a hazard can be controlled or to develop alternatives to a harmful chemical. Algae, daphnia and fish are the most commonly used organisms for environmental toxicity tests, while further tests on other animals may be carried out if necessary. There is a risk that if substances are not broken down they may travel up the food chain, so predators can accumulate large amounts of a particular chemical in their blood when they eat contaminated food. This means that the hazard might be greater to higher animals which are further up the food chain.

Many studies look at the risk of a substance accumulating in the food chain and weigh this against the benefits associated with its use. Birds, such as quails, are often used to study the effects of chemicals on the food chain. The particular species used often represent the native species in the environment being investigated. For example, the Japanese quail is often used to study products used in Japan, whilst the Bob White Quail, which is native to North America, is generally used to test products which are manufactured in the USA. Birds are also convenient for reproductive studies, so the effects of chemicals on development and on successive generations of animals can be investigated.

The Minamata disaster

The disaster at Minamata showed the consequences of environmental pollution and paved the way for protective legislation. Minamata disease was identified in 1956 following many similar incidences nervous system damage in the area local to the Japanese city of Minamata. Pollution was officially recognised to be the cause of the illness in 1968.

Minamata disease was caused by a methyl mercury compound produced by the Chisso Minamata factory. The factory began production of vinyl chloride in 1941, but the official report confirming that chemicals produced by the factory had entered the food chain through the water supply, was not issued by the Japanese government until 27 years later. Since the mercury could not be broken down by the body people who ate fish or shellfish from the Minamata river suffered toxic effects of mercury poisoning.

Since 1968 bioaccumulation and environmental toxicity have been serious concerns governing the regulation of chemical use and testing in Japan. Fish in Minamata Bay were declared safe in 1997, and legislation has been introduced to ensure the prevention of similar disasters. Increased awareness of environmental issues is now leading to the introduction of worldwide regulations and safeguards.

Fluorinated polymers

Fluorinated polymers are water resistant compounds are used in many different chemical products such as grease-proof food packaging, paints and sealants, stain-free carpets and screen-wash fluid. When they degrade chemicals belonging to the PFCA and PFO groups are formed. These cannot be broken down further by bacteria, so they persist in the environment, and are likely to remain in soil for hundreds of years. A team at the University of Toronto led investigations into the prevalence and effects of these chemicals, finding that PFCAs and PFOs are now present in human blood and household dust. There were greater concerns when larger concentrations were discovered in the blood of Arctic mammals such as seals and polar bears, far from any possible source of industrial pollution. These concentrations appeared to be increasing rapidly.

Animal tests have linked PFCAs to cancer and other developmental defects, and they are considered likely to also cause cancer in humans. However, while the possibility of human cancers being caused by these agents presents a hazard, the present risk is thought to be low. The primary cause of concern is that PFCAs appear to be accumulating in the Arctic, far away from any source of chemical pollution. Recent investigations are looking into the effects of these chemicals on rats and fish, while studying how they accumulate in seals and other higher animals.

The researchers are trying to understand how these chemicals move in the environment, and how this leads their accumulation in these remote regions. Since fluorinated polymers have been used since the 1950s an important aspect of this work has been to estimate the total quantity of PFCAs that have been released into the environment. Early assessments showed that the main environmental 'sink' for

PFO and probably of most other PFCAs, was water which carried them to distant regions in rivers, streams and lakes. The research showed that most PFCAs in the environment came directly from the use of chemical products. The breakdown of chemicals into PFCAs appeared to make a smaller, but still important, contribution to pollution. PFCAs are now believed to be carried by both air and water. There is now a focus on reducing emissions of these chemicals during manufacturing processes, and on preventing the release of their precursors into the environment.

The concerns raised by the accumulation of these pollutants in the environment have led to the development of new materials to replace them. Some fluorinated polymer-based products (such as flourotelomers) are not regarded as hazardous and do not appear to have similar toxic effects on the environment. Animal studies are used to verify that they do not metabolise to PFCAs or PFOs, that they are not mutagenic, and that they represent lower risk to the environment. Flourinated polymers are now heavily regulated due to the level of risk they pose the environment. The EPA has introduced a programme requiring manufacturers to reduce their output of toxic impurities due to fluorinated polymers by 2010, while in Europe fluorinated polymers are regulated under REACH to minimise the build-up of PFCAs in the environment.

Endocrine toxicity testing

Endocrine disruption is a potentially serious side effect of chemicals, but there is little call for risk assessment in this area, largely because the science is not sufficiently developed to provide screening tests. Screening of some target chemicals is carried out in the EU (overseen by the endocrine disruption regulation group which works on this problem), but the available tests are currently limited. Under REACH legislation it is usually the responsibility of manufacturers to decide whether chemicals might be endocrine disruptors, but substances already classified as toxic to the reproductive or endocrine systems may be considered 'substances of very high concern', and would need to be tested for these effects to see how they are caused and how they might be prevented.

One way to study effects on the endocrine system is to use reproductive toxicology studies, in which a group of animals are exposed to a chemical before or during pregnancy. Effects of the substance on their ability to reproduce and on successive generations of offspring are studied. However, while these studies may indicate that a chemical has an effect on reproductive hormones, they do not show what the effect is. These tests have recently led to the classification of borates as toxic to reproduction by the European Commission.

Analytical toxicology is the detection, identification, and measurement of foreign compounds (xenobiotics) in biological and other specimens. Analytical methods are available for a very wide range of compounds: these may be chemicals, pesticides, pharmaceuticals, drugs of abuse and natural toxins. Analytical toxicology can assist in the diagnosis, management, prognosis, and prevention of poisoning. In addition analytical toxicology laboratories may be involved in a range of other activities such as the assessment of exposure following chemical incidents, therapeutic drug monitoring, forensic analyses, and monitoring for drugs of abuse. They may also be involved in research, for example in determining the pharmacokinetic and toxicokinetic properties of substances or the efficacy of new treatment regimens. Toxicology is a scientific discipline, overlapping with biology, chemistry, pharmacology, and medicine, that involves the study of the adverse effects of chemical substances on living organisms and the practice of diagnosing and treating exposures to toxins and toxicants. The relationship between dose and its effects on the exposed organism is of high significance in toxicology. Factors that influence chemical toxicity include the dosage, duration of exposure (whether it is acute or chronic), route of exposure, species, age, sex, and environment. Toxicologists are experts on poisons and poisoning.

History of Endocrine toxicity

Dioscorides, a Greek physician in the court of the Roman emperor Nero, made the first attempt to classify plants according to their toxic and therapeutic effect. Ibn Wahshiyya wrote the Book on Poisons in the 9th or 10th century. This was followed up in 1360 by Khagendra Mani Darpana. Mathieu Orfila is considered the modern father of toxicology, having given the subject its first formal treatment in 1813 in his Traité des poisons, also called Toxicologie générale. In 1850, Jean Stas became the first person to successfully isolate plant poisons from human tissue. This allowed him to identify the use of nicotine as a poison in the Bocarmé murder case, providing the evidence needed to convict the Belgian Count Hippolyte Visart de Bocarmé of killing his brother-in-law.

Theophrastus Phillipus Auroleus Bombastus von Hohenheim(also referred to as Paracelsus, from his belief that his studies were above or beyond the work of Celsus – a Roman physician from the first century) is also considered "the father" of toxicology. He is credited with the classic toxicology maxim, "Alle Dinge sind Gift und nichts ist ohne Gift; allein die Dosis macht, dass ein Ding kein Gift ist." which translates as, "All things are poisonous and nothing is without poison; only the dose makes a thing not poisonous." This is often condensed to: "The dose makes the poison" or in Latin "Sola dosis facit venenum".

Basic principles of toxcity assessment

The goal of toxicity assessment is to identify adverse effects of a substance. Adverse effects depend on two main factors: i) routes of exposure (oral, inhalation, or dermal) and ii) dose (duration and concentration of exposure). To explore dose, substances are tested in both acute and chronic models. Generally, different sets of experiments are conducted to determine whether a substance causes cancer and to examine other forms of toxicity.

Factors that influence chemical toxicity:

1. Dosage

- Both large single exposures (acute) and continuous small exposures (chronic) are studied.

2. Route of exposure

- Both
- Ingestion, inhalation or skin absorption

3. Other factors

- Species
- Age
- Sex
- Health
- Environment
- Individual characteristics

TESTING METHODS

Toxicity experiments may be conducted in vivo (using the whole animal) or in vitro (testing on isolated cells or tissues), or in silico (in a computer simulation).

Non-human animals

The classic experimental tool of toxicology is testing on non-human animals. Example of model organisms are Galleria mellonella, which can replace small mammals, and Zebrafish, which allow for the study of toxicology in a lower order vertebrate in vivo. As of 2014, such animal testing provides information that is not available by other means about how substances function in a living organism. The use of non-human animals for toxicology testing is opposed by some organisations for reasons of animal welfare, and it has been restricted or banned under some circumstances in certain regions, such as the testing of cosmetics in the European Union.

Alternative testing methods

While testing in animal models remains as a method of estimating human effects, there are both ethical and technical concerns with animal testing. Since the late 1950s, the field of toxicology has sought to reduce or eliminate animal testing under the rubric of "Three Rs" - reduce the number of experiments with animals to the minimum necessary; refine experiments to cause less suffering, and replace in vivo experiments with other types, or use more simple forms of life when possible. Computer modeling is an example of alternative testing methods; using computer models of chemicals and proteins, structure-activity relationships can be determined, and chemical structures that are likely to bind to, and interfere with, proteins with essential functions, can be identified. This work requires expert knowledge in molecular modeling and statistics together with expert judgment in chemistry, biology and toxicology. In 2007 the American NGO National Academy of Sciences published a report called "Toxicity Testing in the 21st Century:

A Vision and a Strategy" which opened with a statement: "Change often involves a pivotal event that builds on previous history and opens the door to a new era. Pivotal events in science include the discovery of penicillin, the elucidation of the DNA double helix, and the development of computers. Toxicity testing is approaching such a scientific pivot point. It is poised to take advantage of the revolutions in biology and biotechnology. Advances in toxicogenomics, bioinformatics, systems biology, epigenetics, and computational toxicology could transform toxicity testing from a system based on whole-animal testing to one founded primarily on in vitro methods that evaluate changes in biologic processes using cells, cell lines, or cellular components, preferably of human origin." As of 2014 that vision was still unrealized. In some cases shifts away from animal studies has been mandated by law or regulation; the European Union (EU) prohibited use of animal testing for cosmetics in 2013.

Dose response complexities

Most chemicals display a classic dose response curve – at a low dose (below a threshold), no effect is observed. Some show a phenomenon known as sufficient challenge – a small exposure produces animals that "grow more rapidly, have better general appearance and coat quality, have fewer tumors, and live longer than the control animals". A few chemicals have no well-defined safe level of exposure. These are treated with special care. Some chemicals are subject to bioaccumulation as they are stored in rather than being excreted from the body; these also receive special consideration. Several measures are commonly used to describe toxic dosages according to the degree of effect on an organism or a population, and some are specifically defined by various laws or organizational usage. These include:

- LD50 = Median lethal dose, a dose that will kill 50% of an exposed population
- NOEL = No Observed Effect Level, the highest dose known to show no effect
- NOAEL = No Observed Adverse Effect Level, the highest dose known to show no adverse effects
- PEL = Permissible Exposure Limit, the highest concentration permitted under US OSHA regulations
- STEL = Short-Term Exposure Limit, the highest concentration permitted for short periods of time, in general 15–30 minutes
- TWA = Time-Weighted Average, the average amount of an agent's concentration over a specified period of time, usually 8 hours.
- TTC = Threshold of Toxicological Concern have been established for the constituents of tobacco smoke

Types of toxcicology

Medical toxicology

Medical toxicology is the discipline that requires physician status (MD or DO degree plus specialty education and experience).

Clinical toxicology

"Clinical toxicology" redirects here. For the journal, see Clinical Toxicology. Clinical toxicology is the discipline that can be practiced not only by physicians but also other health professionals with a master's degree in clinical toxicology: physician extenders (physician assistants, nurse practitioners), nurses, pharmacists, and allied health professionals.

Computational toxicology

Computational toxicology is a discipline that develops mathematical and computer-based models to better understand and predict adverse health effects caused by chemicals, such as environmental pollutants and pharmaceuticals.

Within the Toxicology in the 21st Century project, the best predictive models were identified to be Deep Neural Networks, Random Forest, and Support Vector Machines, which can reach the performance of in vitro experiments.

Toxicology as a profession

This section does not cite any sources. Please help improve this section by adding citations to reliable sources. Unsourced material may be challenged and removed. (June 2014) A toxicologist is a scientist or medical personnel who specializes in the study of symptoms, mechanisms, treatments and detection of venoms and toxins; especially the poisoning of people.

Requirements of toxicologist

To work as a toxicologist one should obtain a degree in toxicology or a related degree like biology, chemistry, pharmacology or biochemistry. Bachelor's degree programs in toxicology cover the chemical makeup of toxins and their effects on biochemistry, physiology and ecology. After introductory life science courses are complete, students typically enroll in labs and apply toxicology principles to research and other studies. Advanced students delve into specific sectors, like the pharmaceutical industry or law enforcement, which apply methods of toxicology in their work. The Society of Toxicology (SOT) recommends that undergraduates in postsecondary schools that don't offer a bachelor's degree in toxicology consider attaining a degree in biology or chemistry. Additionally, the SOT advises aspiring toxicologists to take statistics and mathematics courses, as well as gain laboratory experience through lab courses, student research projects and internships.

Duties of toxicologist

Toxicologists perform many different duties including research in the academic, nonprofit and industrial fields, product safety evaluation, consulting, public service and legal regulation. In order to research and assess the effects of chemicals, toxicologists perform carefully designed studies and experiments. These experiments help identify the specific amount of a chemical that may cause harm and potential risks of being near or using products that contain certain chemicals. Research projects may range from assessing the effects of toxic pollutants on the environment to evaluating how the human immune system responds to chemical compounds within pharmaceutical drugs. While the basic duties of toxicologists are to determine the effects of chemicals on organisms and their surroundings, specific job duties may vary based on industry and employment. For example, forensic toxicologists may look for toxic substances in a crime scene, whereas aquatic toxicologists may analyze the toxicity level of water bodies.

Scope of Work and Study

Environmental toxicologists work in academia, companies, government agencies, and elsewhere. The work can include laboratory studies, computer modeling, and work "in the field." It is not unusual for an environmental toxicologist to also have training in other areas—for example, public health, environmental chemistry, and pharmacology.

Some examples of what Environmental Toxicologists study include:

- The effects of a chemical or other substance at various concentrations on various species.
- Whether a chemical or other substance can bioaccumulate (increase over time) in animals or other organisms. This is important for human exposures if the bioaccumulation occurs in animals that are part of the human food chain, such as fish.
- Emerging issues such as the study of the sources and effects of microplastics that could become part of the human food chain.

Another emerging global issue is the health of bees. In the news in recent years are terms like the Colony Collapse Disorder (CCD), pesticides like the neonicotinic pesticides (also called neonicotinoids), and parasites that only reproduce in bee colonies. About 75% of all flowering plants rely on animal pollinators and about one-third of our food production is dependent on animal pollinators. The general declining health of honeybees and other bees is thought to be related to complex interactions among multiple stressors, including pesticides and parasites, and stressors like poor nutrition due to declining foraging habitats, bee management practices, and a lack of genetic diversity. Neonicotinoids have been a focus of international attention and their mode of action is on the central nervous system of insects. Neonicotinoids are highly toxic to honeybees and also native bees like bumble bees and blue orchard bees, and sub-lethal levels can affect foraging and the ability to reproduce. Further, neonicotinoids can be persistent in the environment, and can be absorbed by plants and found in pollen and nectar.

Environmental toxicology, field of study in the environmental sciences that is concerned with the assessment of toxic substances in the environment. Although it is based on toxicology, environmental toxicology draws heavily on principles and techniques from other fields, including biochemistry, cell biology, developmental biology, and genetics. Among its primary interests are the assessment of toxic substances in the environment, the monitoring of environments for the presence of toxic substances, the effects of toxins on biotic and abiotic components of ecosystems, and the metabolism and biological and environmental fate of toxins.

HISTORICAL DEVELOPMENT OF ENVIRONMENTAL TOXICOLOGY

Environmental toxicology is a relatively young field, with its origins in the mid-20th century. The modern science of toxicology, on the other hand, was born in the early 19th century, and by the later decades of that century, some scientists had begun to consider the effects of toxic substances that had been released into the environment. But awareness of environmental pollutants did not increase markedly until the publication of American biologist Rachel Carson's Silent Spring in 1962. Despite strong opposition from the chemical industry, which felt that Carson's work unfairly attacked their products, Carson highlighted the environmental side effects from the use of pesticides such as DDT. The book suggested that pollutants used in one area could quickly affect neighbouring areas and that the destruction of a particular part of the food chain upsets the balance of nature, leading to the destruction of an ecosystem. In 1969 scientist René Truhart coined the term ecotoxicology to describe the study of the toxic effects of pollutants on the biological components of ecosystems. Although narrower in scope, ecotoxicology played an important role in the development of environmental toxicology. In the 1970s scientists concerned with toxins in the environment increasingly focused their research on the impacts of agents of biological warfare (e.g., Agent Orange), on industrial pollution, and on discharge from mines.

In the next decade the relevance of environmental toxicology to modern society was realized with the Bhopal disaster, in which 45 tons of poisonous methyl isocyanate gas escaped from an insecticide plant, and the Chernobyl accident, in which massive amounts of radioactive material were released into the atmosphere following the explosion of a nuclear reactor. The Bhopal disaster killed between 15,000 and 20,000 people, with many thousands more suffering from acute and chronic conditions. The Chernobyl accident ultimately was responsible for an estimated several thousand deaths from radiation sickness and cancer and the contamination of millions of acres of land across Belarus, Russia, and Ukraine. In the late 20th and early 21st centuries, the field of environmental toxicology expanded. Among its major concerns were oil spills, the dumping of medical and nuclear waste, air and water pollution, and the impact of substances such as synthetic hormones that were regularly released into environmental reservoirs.

Assessment and monitoring of toxic substances

Toxins affect the environment and organisms in a variety of ways, from having little negative impact on certain abiotic factors or resistant organisms to killing animals and destroying major components of ecosystems. The extent of damage depends on the type and structure of the toxic substance; the age, the size, and the species of the organism; and the temperature and the physical and chemical characteristics of the environment (whether terrestrial or aquatic). Knowledge of how these factors interact is critical to understanding how best to prevent or reduce exposure or remove a toxin from the environment (environmental remediation).

The assessment of toxicity at the levels of whole organism, cell, and gene is one way by which researchers are able to determine how much of a toxin an organism can be exposed to before adverse effects set in. Different assays are used for toxicity assessment, including acute and subacute toxicity assays, sediment toxicity assays, and genotoxicity assays. The determination of safe exposure levels in animals plays a key role in the development of regulations that dictate how toxic substances are to be handled and disposed of. There are also methods by which scientists are able to estimate the quantity of a given toxic substance in the environment. The identification of ways to monitor for chemicals in the environment is an important aspect of environmental toxicology. Monitoring typically is based on the detection of sensitive biochemical markers (e.g., proteins), the levels of which change in the presence of a given toxin, or on changes in individual "indicator" species, the well-being of which serves as a measure of environmental conditions and the health of other species.

Ill Effects of Chemicals on human health

An acute effect of a contaminant (The term "contaminant" means hazardous substances, pollutants, pollution, and chemicals) is one that occurs rapidly after exposure to a large amount of that substance. A chronic effect of a contaminant results from exposure to small amounts of a substance over a long period of time. In such a case, the effect may not be immediately obvious. Chronic effect are difficult to measure, as the effects may not be seen for years. Long-term exposure to cigarette smoking, low level radiation exposure, and moderate alcohol use are all thought to produce chronic effects. For centuries, scientists have known that just about any substance is toxic in sufficient quantities. For example, small amounts of selenium are required by living organisms for proper functioning, but large amounts may cause cancer. The effect of a certain chemical on an individual depends on the dose (amount) of the chemical.

This relationship is often illustrated by a dose-response curve which shows the relationship between dose and the response of the individual. Lethal doses in humans have been determined for many substances from information gathered from records of homicides, accidental poisonings, and testing on animals. A dose that is lethal to 50% of a population of test animals is called the lethal dose-50% or LD-50. Determination of the LD-50 is required for new synthetic chemicals in order to give a measure of their toxicity. A dose that causes 50% of a population to exhibit any significant response (e.g., hair loss, stunted development) is referred to as the effective dose-50% or ED-50. Some toxins have a threshold amount below which there is no apparent effect on the exposed population.

Environmental Contaminants

The contamination of the air, water, or soil with potentially harmful substances can affect any person or community. Contaminants are often chemicals found in the environment in amounts higher than what would be there naturally. We can be exposed to these contaminants from a variety of residential, commercial, and industrial sources. Sometimes harmful environmental contaminants occur biologically, such as mold or a toxic algae bloom.

An Overview of Some Common Contaminants

Arsenic is a naturally occurring element that is normally present throughout our environment in water, soil, dust, air, and food. Levels of arsenic can regionally vary due to farming and industrial activity as well as natural geological processes. The arsenic from farming and smelting tends to bind strongly to soil and is expected to remain near the surface of the land for hundreds of years as a long-term source of exposure. Wood that has been treated with chromated copper arsenate (CCA) is commonly found in decks and railing in existing homes and outdoor structures such as playground equipment. Some underground aquifers are located in rock or soil that has naturally high arsenic content. Most arsenic gets into the body through ingestion of food or water. Arsenic in drinking water is a problem in many countries around the world, including Bangladesh, Chile, China, Vietnam, Taiwan, India, and the United States. Arsenic may also be found in foods, including rice and some fish, where it is present due to uptake from soil and water. It can also enter the body by breathing dust containing arsenic. Researchers are finding that arsenic, even at low levels, can interfere with the body's endocrine system. Arsenic is also a known human carcinogen associated with skin, lung, bladder, kidney, and liver cancer.

Mercury is a naturally occurring metal, a useful chemical in some products, and a potential health risk. Mercury exists in several forms; the types people are usually exposed to are methylmercury and elemental mercury. Elemental mercury at room temperature is a shiny, silver-white liquid which can produce a harmful odorless vapor. Methylmercury, an organic compound, can build up in the bodies of long-living, predatory fish. To keep mercury out of the fish we eat and the air we breathe, it's important to take mercury-containing products to a hazardous waste facility for disposal. Common products sold today that contain small amounts of mercury include fluorescent lights and button-cell batteries.

Although fish and shellfish have many nutritional benefits, consuming large quantities of fish increases a person's exposure to mercury. Pregnant women who eat fish high in mercury on a regular basis run the risk of permanently damaging their developing fetuses. Children born to these mothers may exhibit motor difficulties, sensory problems and cognitive deficits. Figure 1 identifies the typical (average) amounts of mercury in commonly consumed commercial and sport-caught fish. Bisphenol A (BPA) is a chemical synthesized in large quantities for use primarily in the production of polycarbonate plastics and epoxy resins. Polycarbonate plastics have many applications including use in some food and drink packaging, e.g., water and infant bottles, compact discs, impact-resistant safety equipment, and medical devices. Epoxy resins are used as lacquers to coat metal products such as food cans, bottle tops, and water supply pipes. Some dental sealants and composites may also contribute to BPA exposure. The primary source of exposure to BPA for most people is through the diet. Bisphenol A can leach into food from the protective internal epoxy resin coatings of canned foods and from consumer products such as polycarbonate tableware, food storage containers, water bottles, and baby bottles. The degree to which BPA leaches from polycarbonate bottles into liquid may depend more on the temperature of the liquid or bottle, than the age of the container. BPA can also be found in breast milk.

PREVENTION OF EXPOSURE TO BPA?

Some animal studies suggest that infants and children may be the most vulnerable to the effects of BPA. Parents and caregivers, can make the personal choice to reduce exposures of their infants and children to BPA:

- Don't microwave polycarbonate plastic food containers. Polycarbonate is strong and durable, but over time it may break down from over use at high temperatures.
- Plastic containers have recycle codes on the bottom. Some, but not all, plastics that are marked with recycle codes 3 or 7 may be made with BPA.
- Reduce your use of canned foods.

- When possible, opt for glass, porcelain or stainless steel containers, particularly for hot food or liquids.
- Use baby bottles that are BPA free.

Phthalates are a group of synthetic chemicals used to soften and increase the flexibility of plastic and vinyl. Polyvinyl chloride is made softer and more flexible by the addition of phthalates. Phthalates are used in hundreds of consumer products. Phthalates are used in cosmetics and personal care products, including perfume, hair spray, soap, shampoo, nail polish, and skin moisturizers. They are used in consumer products such as flexible plastic and vinyl toys, shower curtains, wallpaper, vinyl miniblinds, food packaging, and plastic wrap. Exposure to low levels of phthalates may come from eating food packaged in plastic that contains phthalates or breathing dust in rooms with vinyl miniblinds, wallpaper, or recently installed flooring that contain phthalates. We can be exposed to phthalates by drinking water that contains phthalates. Phthalates are suspected to be endocrine disruptors.

Lead is a metal that occurs naturally in the rocks and soil of the earth's crust. It is also produced from burning fossil fuels such as coal, oil, gasoline, and natural gas; mining; and manufacturing. Lead has no distinctive taste or smell. The chemical symbol for elemental lead is Pb. Lead is used to produce batteries, pipes, roofing, scientific electronic equipment, military tracking systems, medical devices, and products to shield X-rays and nuclear radiation. It is used in ceramic glazes and crystal glassware. Because of health concerns, lead and lead compounds were banned from house paint in 1978; from solder used on water pipes in 1986; from gasoline in 1995; from solder used on food cans in 1996; and from tin-coated foil on wine bottles in 1996. The U.S. Food and Drug Administration has set a limit on the amount of lead that can be used in ceramics.

Lead and lead compounds are listed as "reasonably anticipated to be a human carcinogen". It can affect almost every organ and system in your body. It can be equally harmful if breathed or swallowed. The part of the body most sensitive to lead exposure is the central nervous system, especially in children, who are more vulnerable to lead poisoning than adults. A child who swallows large amounts of lead can develop brain damage that can cause convulsions and death; the child can also develop blood anemia, kidney damage, colic, and muscle weakness. Repeated low levels of exposure to lead can alter a child's normal mental and physical growth and result in learning or behavioral problems. Exposure to high levels of lead for pregnant women can cause miscarriage, premature births, and smaller babies.

Repeated or chronic exposure can cause lead to accumulate in your body, leading to lead poisoning.

Formaldehyde is a colorless, flammable gas or liquid that has a pungent, suffocating odor. It is a volatile organic compound, which is an organic compound that easily becomes a vapor or gas. It is also naturally produced in small, harmless amounts in the human body. The primary way we can be exposed to formaldehyde is by breathing air containing it. Releases of formaldehyde into the air occur from industries using or manufacturing formaldehyde, wood products (such as particle-board, plywood, and furniture), automobile exhaust, cigarette smoke, paints and varnishes, and carpets and permanent press fabrics. Nail polish, and commercially applied floor finish emit formaldehyde. In general, indoor environments consistently have higher concentrations than outdoor environments, because many building materials, consumer products, and fabrics emit formaldehyde. Levels of formaldehyde measured in indoor air range from 0.02–4 parts per million (ppm). Formaldehyde levels in outdoor air range from 0.001 to 0.02 ppm in urban areas.

Radiation

Radiation is energy given off by atoms and is all around us. We are exposed to radiation every day from natural sources like soil, rocks, and the sun. We are also exposed to radiation from man-made sources like medical X-rays and smoke detectors. We're even exposed to low levels of radiation on cross-country flights, from watching television, and even from some construction materials. You cannot see, smell or taste radiation. Some types of radioactive materials are more dangerous than others. So it's important to carefully manage radiation and radioactive substances to protect health and the environment.

Radon is a radioactive gas that is naturally-occurring, colorless, and odorless. It comes from the natural decay of uranium or thorium found in nearly all soils. It typically moves up through the ground and into the home through cracks in floors, walls and foundations. It can also be released from building materials or from well water. Radon breaks down quickly, giving off radioactive particles. Long-term exposure to these particles can lead to lung cancer. Radon is the leading cause of lung cancer among nonsmokers, according to the U.S. Environmental Protection Agency, and the second leading cause behind smoking.

What is NIEHS Doing?

Supporting grantees across the country. NIEHS has a long history of funding grants in toxicology and environmental health sciences. For example researchers supported through the NIEHS Division of Extramural Research and Training are working to develop new and improved models of toxicity that can help predict cancer and other adverse health outcomes that may result from fetal or early life exposures. For example, because we know that people are exposed to a multitude of environmental toxicants, NIEHS grantees are working to understand the implications of chemical mixtures, which can sometimes produce health affects greater than each chemical would alone.

Grantees have published some of the first immunotoxic studies linking childhood exposures to perfluorinated compounds (PFCs) with immune system deficiency. NIEHS supported researchers have also used their toxicology expertise to improve our knowledge and our approaches to chemicals such as arsenic, leady and mercury. Also, grantees are active participants in our nanotechnology efforts, where we are working to increase our understanding of how engineered nanomaterials interact with living systems, so we can develop predictive models for quantifying exposure from nanomaterials and are able to assess their health impacts. These efforts will help guide the design of second-generation engineered nanomaterials and minimize adverse health effects.

Also, the NIEHS Superfund Research Program works to gain a better understanding of how toxicants affect human health in order to help environmental managers and risk assessors protect the public from exposures to hazardous substances. They accomplish this through research conducted at universities across the country, to develop cost-effective approaches to detect, remove, and/or reduce the amount of toxic substances found in the environment. The program conducts research on remediation, detection, and monitoring tools or strategies and also funds studies that relate to multiple disease endpoints such as cardiovascular disease, cancer, and neurological disorders. They may also investigate the relationship between dose and effect, which can be influenced by factors such as the dosing regimen (single large exposure vs. continuous smaller exposures), route of exposure (oral, dermal, nasal), age, gender, and environment. Toxicology brings together a wide variety of fields, including chemistry, biology, pharmacology, human and animal medicine, and environmental science, to help inform policies and regulations to protect both human health and the environment. Toxicologists spend their time planning and conducting experiments, dosing animals, and collecting and analyzing data. Ph.D. level toxicologists interpret the results of studies, conduct risk assessments, and integrate data from many different studies. They then create reports and recommendations for organizations and regulatory agencies, putting the data into context and providing risk analyses to ensure the safety of the products and compounds for their intended use. Toxicologists study the safety and biological effects of drugs, chemicals, agents, and other substances on living organisms. They develop methods to determine harmful effects, the dosages that cause those effects, and safe exposure limits.

Ecotoxicology is the study of the effects of toxic chemicals on biological organisms, especially at the population, community, ecosystem, and biosphere levels. Ecotoxicology is a multidisciplinary field, which integrates toxicology and ecology. The ultimate goal of this approach is to be able to reveal and to predict the effects of pollution within the context of all other environmental factors. Based on this knowledge the most efficient and effective action to prevent or remediate any detrimental effect can be identified. In those ecosystems that are already impacted by pollution ecotoxicological studies can inform as to the best course of action to restore ecosystem services and functions efficiently and effectively. Ecotoxicology differs from environmental toxicology in that it integrates the effects of stressors across all levels of biological organisation from the molecular to whole communities and ecosystems, whereas environmental toxicology focuses upon effects at the level of the individual and below.

Common environmental toxicants

- PCBs (polychlorinated biphenyls) - found in coolant and insulating fluids, pesticide extenders, adhesives, and hydraulic fluids.
- Pesticides - used widely for preventing, destroying, or repelling any organism that may be considered harmful. Commonly found in commercially grown fruits, vegetables, and meats. Methyl parathion is a commonly used pesticide used for agricultural reasons. Methyl parathion causes the formation of toxic mediums for humans, soil and water, fresh water fish, and other hydrophilous organisms in the ecosystem. Methyl parathion proposes numerous health risk factors that are life-threatening.
- Mold and other mycotoxins.
- Phthalates are found in plastic wrap, plastic bottles, and plastic food storage containers, all of which make up a considerable part of household plastic waste.
- VOCs (volatile organic compounds) - such as formaldehyde; can be found in drinking water and sewage systems.
- Dioxins are a class of chemical compounds that are formed as a result of combustion processes such as waste incineration and from burning fuels like wood, coal, and oil.

- Asbestos is found in the insulation of flows, ceilings, water pipes, and heating ducts.
- Heavy metals include arsenic, mercury, lead, aluminum, and cadmium, which are found in fish, and pesticides.
- Chloroform is used to make other chemicals.
- Chlorine is commonly found in household cleaners.

Exposure to toxic chemicals

- Chemicals propose the risk of killing off another animal's food supply that changes the overall population of the prey.
- Animals can go to the brink of extinction because of the food chain that exists through the different communities. For example, bald eagles, ospreys, and peregrine falcons were facing extinction because their food sources(fish and other birds) were contaminated with toxins.
- We are all connected between the communities of living things. Plants can absorb toxins through their roots and leaves. Animals and humans are always exposed to chemicals by the air we breathe, things we touch, and what we put in our mouth.
- Animals and humans can also eat other animals or plants that are already poisoned, which will continue the spread of chemicals, which is referred to as secondary poisoning.

3

TOXICITY TESTING

Toxicology testing, also known as safety testing, or toxicity testing, is conducted to determine the degree to which a substance can damage a living or non-living organisms. It is often conducted by researchers using standard test procedures to comply with governing regulations. Some toxicology testing is chemical testing while some others use animals in the laboratory studies. Contract animal testing facilities such as Huntingdon Life Sciences and Inveresk Research International conduct a variety of tests on behalf of customers, including the manufacturer of medicines and household products.

HISTORY

Europe

Around one million animals, primate and non–primate, are used every year in Europe in toxicology tests. In the UK, one-fifth of animal experiments are toxicology tests.

Methodology

Toxicity tests examine finished products such as pesticides, medications, food additives such as artificial sweeteners, packing materials, and air freshener, or their chemical ingredients. The substances are applied to the skin or eyes; injected intravenously, intramuscularly, or subcutaneously; inhaled either by placing a mask over the animals and restraining them, or by placing them in an inhalation chamber; or administered orally, through a tube into the stomach, or placing them in the animals' food. Doses may be given once, repeated regularly for many months, or for the lifespan of the animal. Toxicity tests can also be conducted on materials need to be disposed such as sediment to be disposed in marine environment.

Contract Research Organizations

A Contract Research Organization (CRO) is an organization that provides support to the pharmaceutical, biotechnology, chemical, and medical device industries in the form of research services outsourced on a contract basis. A CRO may provide toxicity testing services, along with others such as assay development, preclinical research, clinical research, clinical trials management, and pharmacovigilance. CROs also support foundations, research institutions, and universities, in addition to governmental organizations (such as the NIH, EMEA, etc.).

Regulation

United States

In the United States, toxicology tests are subject to Good Laboratory Practice guidelines and other Food and Drug Administration laws.

Toxicity Tests

A test method is a definitive procedure that produces a test result. A toxicity test, by extension, is designed to generate data concerning the adverse effects of a substance on human or animal health, or the environment. Many toxicity tests examine specific types of adverse effects, known as endpoints, such as eye irritation or cancer. Other tests are more general in nature, ranging from acute (single-exposure) studies to repeat dose (multiple-exposure) studies, in which animals are administered daily doses of a test substance.

Toxicity endpoints considered within the scope of AltTox include the following:

Acute Systemic Toxicity

Adverse effects occurring within a relatively short time after administration of a single (typically high) dose of a substance via one or more of the following exposure routes: oral, inhalation, skin, or injection

Biologics, Vaccines, Medical Products

Variety of tests and endpoints related to efficacy, safety, and quality control testing of drugs, devices, biologics, diagnostics, and vaccines (of considerable interest for the development of non-animal methods due to the large numbers of animals used).

Carcinogenicity

Chemically-induced cancer, whether through genotoxic or non-genotoxic (e.g., growth-promoting) mechanisms

Dermal Penetration

Extent and rate by which a chemical is able to enter the body via the skin; also known as skin or percutaneous absorption.

Ecotoxicity

Chemically-induced adverse effects on organisms in the environment, including mammals, birds, fish, amphibians, crustaceans, other aquatic invertebrates, and even plants; common study designs include acute systemic, dietary, and reproductive (also known as life-cycle) toxicity

Endocrine Disruptors

Substances that interact with the hormonal systems of humans and/or wildlife, and thereby disrupt normal biological functions

Eye Irritation/Corrosion

Chemically-induced eye damage that is reversible (irritation) or irreversible (corrosion)

Genotoxicity

Chemically-induced mutations and/or other alterations in the structure, information content, or segregation of genetic material (e.g., DNA strand breaks or a gain/loss in chromosome number)

Neurotoxicity

Chemically-induced adverse effects on the brain, spinal cord, and/or peripheral nervous system (e.g., deficits in learning or sensory ability)

Pharmacokinetics and Metabolism

Study of the absorption, distribution, metabolism, and elimination (ADME) of drugs or chemicals in the body; also known as toxicokinetics:

Phototoxicity

Toxic responses from a substance (applied to the body or ingested) following exposure to light or skin irradiation

Repeated Dose/Organ Toxicity

General toxicological effects occurring as a result of repeated daily exposure to a substance (via oral, inhalation, dermal, or injection routes) for a portion of the expected life span (i.e., subacute or subchronic exposure), or for the majority of the life span (i.e., chronic exposure)

REPRODUCTIVE AND DEVELOPMENTAL TOXICITY

Chemically-induced adverse effects on sexual function, fertility, and/or normal offspring development (e.g., spontaneous abortion, premature delivery, or birth defects); generally determined through the breeding of one or more generations of offspring

Skin Irritation/Corrosion

Chemically-induced skin damage that is reversible (irritation) or irreversible (corrosion)

Skin Sensitization

The induction of allergic contact dermatitis following exposure to a chemical agent.

There is a long history in the use of animals as models for toxicity testing. During the 20th century, as regulatory agencies were established by national governments, animal test guidelines were added to address regulatory requirements that products need to be "safe" for consumers. A review on pre-clinical testing by Parasuraman (2011) further explains the animal models and tests used for regulatory toxicity testing.

There are many advantages, including scientific, ethical, and economic ones, for replacing the animal toxicity tests with non-animal (in vitro and in silico) test systems. As science progresses, we have become aware of a number of problems with the use of animal models for assessing the safety of chemicals, drugs, and other products to human health, some of which are briefly discussed in the following paragraphs.

Product Safety Testing Methods have not Kept Pace with Scientific Progress

Between the time that most commonly used toxicity tests were conceived and today, there has been a revolution in biology and biotechnology. Advances in cell culture and robotics have given birth to rapid "high throughput" in vitro test systems, while tissue engineering and stem cell technologies are providing ever more relevant in vitro tissues. Emerging technologies such as genomics, proteomics, metabolomics, computational biology, and in silico (computer-based) systems offer still more potential alternatives to animal test methods. The 2007 National Academy of Sciences report, Toxicity testing in the twenty-first century, a vision and a strategy, called for a major paradigm shift in toxicology that would "rely less heavily on animal studies and instead focus on in vitro methods that evaluate chemicals' effects on biological processes using cells, cell lines, or cellular components, preferably of human origin. The new approach would generate more-relevant data to evaluate risks people face, expand the number of chemicals that could be scrutinized, and reduce the time, money, and animals involved in testing."

Integrated testing strategies based on combinations of advanced in vitro and in silico methods that model the mechanism(s) of action and address the adverse outcome pathway are now the focus of research and development in several prototype initiatives.

Questionable Reliability and Relevance of Animal Methods/Models

Animal testing is predicated on the assumption that adverse effects observed in one animal species could also occur in others. However, it is widely recognized within the scientific community that different species can respond differently to the same substance. Whether interspecies differences are due to genetic, biochemical, or metabolic factors—or a combination—the results of testing on rodents, rabbits, or dogs may not provide an accurate prediction of toxic effects in humans. It has also been recognized that the health status, genetic features, developmental stage, and other factors impact an individual's response to chemical exposure, and this complex interplay is extremely difficult to account for in a test with standardized, inbred animals.

There has also been a concern about the relevance of extrapolating from the high doses of test substances commonly administered to animals to realistic human or environmental exposure levels. High doses commonly used in some animal tests may pose another problem—frank or overwhelming toxicity that causes the animals to exhibit stress responses, making results difficult to interpret. This has been a problem particularly for tests focusing on reproductive, developmental, and carcinogenicity endpoints. In addition, despite efforts to standardize procedures, the results of some animal tests can be highly variable and difficult to reproduce.

Animal Welfare Considerations

Some conventional toxicity test methods consume hundreds to thousands of animals per substance examined. Statistics on animal use, such as the Sixth Report on the Statistics on the Number of Animals used for Experimental and Other Scientific Purposes in the Member States of the European Union (2010), indicate that toxicity testing accounts for a large percentage of the more painful procedures experienced by animals (e.g., the use of death as the experimental endpoint in acute systemic toxicity studies).

Large government testing program requirements, such as REACH and HPV, greatly increased the animal testing already conducted by companies. While REACH requires that animal testing is used "only when there are no other scientifically reliable ways of assessing the potential effects on humans or the environment," the latest report by the European Chemicals Agency on the use of alternatives in testing for the REACH regulation explains that companies are not fully implementing the use of animal alternative methods, and that animal tests continue to increase due to new registrations (ECHA, 2014).

As public opposition towards animal testing has grown, some parts of the world have broadly prohibited testing on animals where alternative methods are "reasonably and practicably available" (e.g., EU Directive 2010/63/EU on the protection of animals used for scientific purposes). Animal testing bans may also be sector-specific, as in the case of the EU Cosmetics Regulation (EC) No 1223/2009, which now bans the marketing of any cosmetic product within the EU containing ingredients that have been tested on animals. Earlier EU test bans on cosmetics ingredients and final formulations were also imposed under the 7th Amendment to the EU Cosmetics Directive (76/768/EEC). The EU's 6th animals used statistics reported a decrease in animal use for toxicity testing of industry and agricultural products, and environment testing, and attributed this to implementation of laws phasing out animal testing for cosmetics in the EU. The ECHA report, The Use of Alternatives to Testing on Animals for the REACH Regulation (2014), also reported that the use animal alternative methods had increased (although not fully implemented), and, therefore, did reduce the overall numbers of animals that would have otherwise been used.

Time and Cost Considerations

Some conventional tests take months to years to conduct and analyze (e.g., 4-5 years for carcinogenicity studies), at a cost of hundreds of thousands to millions of dollars per substance examined (e.g., $2-4 million per two-species carcinogenicity study).

In addition to the testing required for new chemicals and products, there is growing realization that the safety information we have on many existing chemicals is insufficient. While it is desirable to have only safe chemicals and products in the marketplace, the traditional means for conducting hazard assessments cannot keep pace with the demand. This has been the basis of a number of government testing programs such as EU's REACH (Registration, Evaluation, Authorisation and Restriction of Chemicals) legislation, and government research programs such as the US EPA's ToxCast™ program.

Non-animal Toxicity Test Methods

AltTox.org is specifically interested in animal replacement methods used for toxicity testing. In cases where replacement methods are not available, reduction and refinement methods are covered.

The term alternative, in the context of toxicity testing, is used to describe any change from present procedures that will result in the replacement of animals, a reduction in the numbers used, or a refinement of techniques to alleviate or minimize pain, distress, and/or suffering. This 3Rs concept of alternatives is rooted in the 1959 publication of Russell and Burch, The Principles of Humane Experimental Technique. Since the publication of this seminal work, governments, industry, NGOs, and other stakeholders have invested substantial time and financial resources to advance the 3Rs in research and testing.

Non-animal methods for producing toxicological data include the following:

In vitro cell and tissue culture-based methods and models.

Toxicity assays can be conducted using models developed using primary cells, cell lines, stem cells, 3-dimensional cultured cells, excised tissues, or cultured organs. Some cell and tissue-based methods have already achieved validation and international acceptance.

In silicosystems.

Computer-based methods such as (quantitative) structure-activity relationship ((Q)SAR) models and read-across can be used to predict the biological/toxicological properties of a substance.

INTEGRATED TESTING AND OTHER EMERGING STRATEGIES

Integrated testing strategies combine methods, such as in silico methods and multiple in vitro assays, along with appropriate statistical analysis, for the prediction of in vivo toxicity responses. An understanding of the mechanisms of toxicity and the cellular pathways involved are currently being investigated with the goal of obtaining more predictive integrated test strategies.

Other strategies for reducing animal testing requirements include:

- Human epidemiology or accidental exposures
- Human volunteer studies

- Existing toxicological information on a substance is recognized as sufficient for risk assessment purposes
- Waiving of a requirement to conduct new testing because human exposure levels are below what is considered a significant risk to human health, or because testing would be difficult, impossible, or meaningless given the nature of the substance in question

Scientific Validation, Regulatory Acceptance, International Harmonization

To facilitate the replacement of old tests with new ones, international authorities developed processes and criteria for evaluating new toxicity test methods to determine whether they can replace an existing method.

Test method validation refers to the process of determining whether a toxicity test method (or test battery or test scheme) is reliable (reproducible) and relevant for its intended purpose. Relevance in this setting means "the extent to which the test method correctly measures or predicts the (biological) effect of interest."

Criteria and processes for toxicity test method validation were developed in the mid-1990's by validation authorities in the EU and US, ECVAM (European Centre for the Validation of Alternative Methods) and ICCVAM (Interagency Coordinating Committee on the Validation of Alternative Methods), respectively, and by the international Organisation for Economic Cooperation and Development (OECD). By 2005, the validation criteria of these organizations was harmonized and published as the Guidance Document on the Validation and International Acceptance of New or Updated Test Methods for Hazard Assessment (OECD, 2005).

Key steps in the pathway from test method development to widespread use and acceptance of a new toxicity test method are the following – in their typical chronological order:

- Research & development consists of the basic and applied research involved in the genesis of a new test method. It might be undertaken and/or funded by biotech companies, contract testing laboratories, research institutes, regulated industry, or government bodies.
- Prevalidation is the process that aims to establish the mechanistic basis of a test, standardize and optimize the test protocol, evaluate within-lab variability, and define a model or procedure by which test results are used to predict the in vivo toxicological endpoint.

- Validation is the process to evaluate a defined set of criteria that includes a test method's transferability to other laboratories, between-lab variability and reproducibility, and reliability.
- Peer review or independent assessment is the formal evaluation process of the results from a validation study. Recommendations on the validity of the test method for the indicated purpose (as well as other aspects of the study) are made to the sponsoring validation authority.
- Regulatory acceptance processes differ region by region. In Europe, ESAC endorsement can lead to EU-wide acceptance, considering the legal requirement under Directive 86/609/EEC that non-animal alternatives be used preferentially. In the US, agencies individually decide whether to accept an ICCVAM recommendation. Most government agencies have the authority to accept whatever methods they find valid to meet their particular needs.
- International harmonization involves collaborative efforts through established organizations such as the OECD, ICH, and VICH to adopt the same testing requirements and test protocols, so that different countries and regions require and accept the same types of data for regulatory decisions. International acceptance in the form of an OECD Test Guideline is an important milestone for a validated non-animal test method, and can have a significant impact on reducing animal use in regulatory assessments.

Clinical Investigations

Knowledge of toxicity of xenobiotics to humans is derived by three methods:

Clinical Investigations	Administration of Chemicals to human subjects with careful clinical observations and laboratory measurements
Epidemiological studies	Observations of humans that have been exposed to xenobiotics in the normal course of their life or occupation
Adverse reactions to drug reports	Reports voluntarily submitted by physicians to the FDA after a drug has been approved and is in widespread use

Clinical investigations are a component of the Investigational New Drug Applications (IND) submitted to U.S. Food and Drug Administration (FDA).

Clinical investigations are conducted only after the non-clinical laboratory studies have been completed.

Toxicity studies using human subjects require strict ethical considerations. They are primarily conducted for new pharmaceutical applications submitted to FDA for approval.

Generally, toxicity found in animal studies occurs with similar incidence and severity in humans. Differences sometimes occur, thus clinical tests with humans are needed to confirm the results of non-clinical laboratory studies.

FDA clinical investigations are conducted in three phases. Phase 1 consists of testing the drug in a small group of 20-80 patients. Information obtained in Phase 1 studies is used to design Phase 2 studies. In particular, to:

- Determine the drug's pharmacokinetics and pharmacological effects
- Elucidate its metabolism
- Study the mechanism of action of the drug

Phase 2 studies are more extensive involving several hundred patients and are used to:

- Determine the short-term side effects of the drug
- Determine the risks associated with the drug
- Evaluate the effectiveness of the drug for treatment of a particular disease or condition

Phase 3 studies are expanded controlled and uncontrolled trials conducted with several hundred to several thousand patients. They are designed to:

- Gather additional information about effectiveness and safety
- Evaluate overall benefit-risk relationship of the drug
- Provide the basis for the precautionary information that accompanies the drug

Epidemiology Studies

Epidemiology studies are conducted using human populations to evaluate whether there is a causal relationship between exposure to a substance and adverse health effects. These studies differ from clinical investigations in that individuals have already been administered the drug during medical treatment or have been exposed to it in the workplace or environment.

Epidemiological studies measure the risk of illness or death in an exposed population compared to that risk in an identical (e.g., same age, sex, race, social status, etc.), unexposed population.

There are four primary types of epidemiology studies. They are:

Cohort Studies	A cohort (group) of individuals with exposure to a chemical and a cohort without exposure are followed over time to compare disease occurrence.
Case Control Studies	Individuals with a disease (e.g., cancer) are compared with similar individuals without the disease to determine if there is an association of the disease with prior exposure to an agent.
Cross-Sectional Studies	The prevalence of a disease or clinical parameter among one or more exposed groups is studied. For example, the prevalence of respiratory conditions among furniture makers.
Ecological Studies	The incidence of a diseases in one geographical area is compared to that of another area. For example, cancer mortality in areas with hazardous waste sites as compared to areas without waste sites.

Cohort studies are the most commonly conducted epidemiology studies. They frequently involve occupational exposures. Exposed persons are easy to identify and the exposure levels are usually higher than in the general public. There are two types of cohort studies:

Prospective cohort sudy	Cohorts are identified based on current exposures and followed into the future.
Retrospective Cohort study	Cohorts are identified based on past exposure conditions and follow-up proceeds forward in time.
To determine if epidemiological data are meaningful, standard, quantitative measures of effect are employed. The most commonly used are:	
Odds Ratio (O/R)	The ratio of risk of disease in a case-control study for an exposed group to an unexposed group. An O/R=2 means that the exposed group has twice the risk as the non-exposed group.

Standardized Mortality Ratio (SMR)	The relative risk of death based on a comparison of an exposed group to non-exposed group. An SMR=150 indicates that there is a 50% greater risk.
Relative Risk (RR)	The ratio expressing the occurrence of disease in an exposed population to that of an unexposed population. An RR=175 indicates a 75% increase in risk.

There are a number of aspects in designing an epidemiology study. The most critical are appropriate controls, adequate time span, and statistical ability to detect an effect.

The control population used as a comparison group must be as similar as possible to that of the test group, e.g., same age, sex, race, social status, geographical area, and environmental and lifestyle influences.

Many epidemiology studies evaluate the potential for an agent to cause cancer. Since most cancers require long latency periods, e.g., 20 years, the study must cover that period of time.

The statistical ability to detect an effect is referred to as the power of the study. To gain precision, the study and control populations should be as large as possible.

Epidemiologists attempt to control errors that may occur in the collection of data. These errors, known as bias errors, are of three main types:

Selection bias	Occurs when the study group is not representative the population from which it came.
Information bias	Occurs when study subjects are misclassified as to disease or exposure status. Recall bias occurs when individuals are asked to remember exposures or conditions that existed years before.
Confounding factors	Occur when the study and control populations differ with respect to factors which might influence the occurrence of the disease.
Mortality Ratio (SMR)	For example, smoking might be a confounding factor and should be considered when designing studies.

Animal Testing for Toxicity

Animal tests for toxicity are conducted prior to human clinical investigations as part of the non-clinical laboratory tests of pharmaceuticals. For pesticides and industrial chemicals, human testing is rarely conducted. Animal test results often represent the only means by which toxicity in humans can be effectively predicted.

With animal tests:

- Chemical exposure can be precisely controlled
- Environmental conditions can be well-controlled
- Virtually any type of toxic effect can be evaluated
- The mechanism by which toxicity occurs can be studied

Methods to evaluate toxicity exist for a wide variety of toxic effects. Some procedures for routine safety testing have been standardized. Standardized animal toxicity tests are highly effective in detecting toxicity that may occur in humans. Concern for animal welfare has resulted in tests that use humane procedures and only the number of animals needed for statistical reliability.

To be standardized, a test procedure must have scientific acceptance as the most meaningful assay for the toxic effect. Toxicity testing can be very specific for a particular effect, such as dermal irritation, or it may be general, such as testing for unknown chronic effects.

Standardized tests have been developed for the following effects:

- Acute Toxicity
- Subchronic Toxicity
- Chronic Toxicity
- Carcinogenicity
- Reproductive Toxicity
- Developmental Toxicity
- Dermal Toxicity
- Ocular Toxicity
- Neurotoxicity
- Genetic Toxicity

Species selection varies with the toxicity test to be performed. There is no single species of animal that can be used for all toxicity tests. Different species may be needed to assess different types of toxicity. In some cases, it may not be possible to use the most desirable animal for testing because of animal welfare or cost considerations. For example, use of monkeys and dogs is restricted to special cases, even though they represent the species that may react closest to humans.

Rodents and rabbits are the most commonly used laboratory species due to their availability, low costs in breeding and housing, and past history in producing reliable results.

The toxicologist attempts to design an experiment to duplicate the potential exposure of humans as closely as possible. For example:

- The route of exposure should simulate that of human exposure. Most standard tests use inhalation, oral, or dermal routes of exposure.
- The age of test animals should relate to that of humans. Testing is normally conducted with young adults, although newborn or pregnant animals may be used in some cases.
- For most routine tests, both sexes are used. Sex differences in toxic response are minimal, except for toxic substances with hormonal properties.
- Dose levels are normally selected so as to determine the threshold as well as dose-response relationship. Usually, a minimum of three dose levels are used.

Acute Toxicity

Acute toxicity tests are generally the first tests conducted. They provide data on the relative toxicity likely to arise from a single or brief exposure. Standardized tests are available for oral, dermal, and inhalation exposures. Basic parameters of these tests are:

Species	rats preferred for oral and inhalation test; rabbits preferred for dermal tests.
Age	young adults.
Number of animals	5 of each sex per dose level.
Dosage	theee dose levels recommended; exposures and single doses or fractionated dose up to 24 hours for oral and dermal studies; and 4-hour exposure for inhalation studies.
Observation period	14 Days.

Subchronic Toxicity

Subchronic toxicity tests are employed to determine toxicity likely to arise from repeated exposures of several weeks to several months. Standardized tests are available for oral, dermal, and inhalation exposures. Detailed clinical observations and pathology examinations are conducted. Basic parameters of these tests are:

Species	Rodents (useually rats) preferred for oral and inhalation studies; rabbhits for drmal studies; non-rodents (usually dogs) recommended as a second species for oral tests.
Age	Young adults 10of each sex for rodents, 4 of each sex for non-rodents per dose level.
Dosage	Three dose levels plus a control group; include a toxic dose level plus NOAEL; exposures are 90 days.
Observation period	90 days (same as treatment period)

Chronic Toxicity

Chronic toxicity tests determine toxicity from exposure for a substantial portion of a subject's life. They are similar to the subchronic tests except that they extend over a longer period of time and involve larger groups of animals. Basic parameters of these tests include:

Species	Two species recommended; rodent and non-rodent (rat and dog).
Age	young adults.
Numbers of animals	20 of each sex for rodents, 4 of each sex for non-rodents per dose level.
Dosage	three dose levels recommended; includes a toxic dose level and NOAEL; exposures generally for 12 months; FDA requests 24 months for food chemicals.
Observation Period	12-24 months.

Carcinogenicity

Carcinogenicity tests are similar to chronic toxicity tests. However, they extend over a longer period of time and require larger groups of animals in order to assess the potential for cancer. Basic parameters of these tests are:

Species	Testing in two rodents species, the rat and mouse preferred due to relatively short life spans
Age	young adults
Number of animals	50 of each sex per dose level

Dosage	Three dose levels recommended; highest should produce minial toxicity; exposure periods are at least 18 months for mice and 24 months for rats.
Observation Period	18-24 months for mice and 24-30 months for rats.

Reproductive Toxicity

Reproductive toxicity testing is intended to determine the effects of substances on gonadal function, conception, birth, and the growth and development of the offspring. The oral route is preferred. Basic parameters of these tests are:

Species	rat is recommended
Age	Young adults
Number of animals	20 of each sex per dose level
Dosage	Three dose levels recommended; highest dose should produce toxicity but not morality in parents; lowest dose should not produce toxicity.
Observation Period	Test substance given to parental animals (P1) prior to mating, during pregnancy and through weaning of F1 offspring; substance then given to selected F1 offspring during their growth into adulthood, mating, and production of an F2 generation, until the F2 generation is 21 days old.

Developmental Toxicity

Developmental toxicity testing detects the potential for substances to produce embryotoxicity and birth defects. Basic parameters of this test are:

Species	Two species recommended: rat, mouse, hamster, and rabbit most commonly used.
Age	Young adults females
Number of animals	20 pregnant females per dose level.
Dosage	At least three dose levels are used: throughout organ developement in the fetus for teratogenic effects; stars with parents prior to breeding, continues during pregnancy for all devlopmental effects.

Observation Period	Offspring sacrificed and examined day prior to expected birth for reratogenic effects; offspring observed for growth retardation and abnormal function through infancy and examined for teratogenic effects.

Dermal Toxicity

Dermal toxicity tests determine the potential for an agent to cause irritation and inflammation of the skin. This may be the result of direct damage to the skin cells by a substance. It may also be an indirect response due to sensitization from prior exposure. There are two dermal toxicity tests:

Primary Dermal Irritation	Dermines direct toxicity. The substance is applied to the skin of 6 albino rabbits for 4-hours and the rabbits are observed for 72 hours for irritation.
Dermal Sensitization	Assays for immune hypersensitivity of the skin. It consists of two phases. In the sensitization phase the test substance is applied to the skin of guinea pigs for 4 hours. There should be no skin inflammation. At least one week later, the substance is reapplied to the skin. This is known as the challenge phase. An inflammatory reaction indicates that the skin has been sensitized to the substance.

Ocular Toxicity

Ocular toxicity is determined by applying a test substance for one second to the eyes of 6 test animals, usually rabbits. The eyes are then carefully examined for 72-hours, using a magnifying instrument to detect minor effects. The ocular reaction may occur on the cornea, conjunctiva, or iris. It may be simple irritation that is reversible and quickly disappears or the irritation may be severe and produce corrosion, an irreversible condition.

The eye irritation test is commonly known as the "Draize Test." This test has been targeted by animal welfare groups as an inhumane procedure due to pain that may be induced in the eye. The test allows the use of an eye anesthetic in the event pain is evident. The Draize Test is a reliable predictor of human eye response. However, research to develop alternative testing procedures that do not use live animals is underway. While some cell and tissue assays are promising, they have not as yet proved as reliable as the animal test.

Neurotoxicity

A battery of standardized neurotoxicity tests has recently been developed to supplement the delayed neurotoxicity test in domestic chickens (hens). The hen assay determines delayed neurotoxicity resulting from exposure to anticholinergic substances, such as certain pesticides. The hens are protected from the immediate neurological effects of the test substance and observed for 21 days for delayed neurotoxicity. Other neurotoxicity tests include measurements of:

Motor Activity	Tests for decreased motor activity, such as cage movement. Rats or mice are used.
Peripheral Never Conduction	Tests for electrical conduction in motor and sensory nerves. Rodents are exposed for 90 days to the test substance.
Neuropathology	Tests for nerve damage by microscopic examination. This is one aspect of other standardized toxicity tests.

Genetic Toxicity

Genetic toxicity is determined using a wide range of test species including whole animals and plants (e.g., rodents, insects, and corn), microorganisms, and mammalian cells. A large variety of tests have been developed to measure gene mutations, chromosome changes, and DNA activity. The most common gene mutation tests involve:

Microrganisms	Salmonella typhimurium and E. Coll are Commonly used baterial tests. The S. typhimurium assay is known as the Ames Test. yeast are also used to detect gene mulations in culture systems.
Mammalian Cells	The two main cell lines are mouse lymphoma and Chinesse hamster ovary (CHO) Cells.
Fruit Flies	Drosophila melanogaster is used to detect sex-linked recessive lethal mutations.
Mice	The mouse Specific Locus TEst is the major gene mutation test that employs whole animals. Exposed mice are bred and observed for hereditary changes.

Chromosomal effects can be detected by a variety of tests, some involving whole animals (in vivo). Some use cell systems (in vitro). Several assays are available to test for chemically induced chromosome aberrations in whole animals. The most common tests are:

Rodent Chromosomal Assay (in vivo)	Involves exposure of mice or rats to a single dose of a substance. Their bone marrow is analyzed for chromosome aberrratios over a 48 hour period.
Dominant Lethal	Exposed male mice or rats are mated with untereated females.
Assay	The presence of dead implants or fetuses is the result of the fertilized ovum receiving damaged DNA from the sperm.
Micronucles Test	Mice are exposed once and their bone marrow or peripheral blood cells are examined for 72 hours for the presence of micronuclei, such as broken pieces of chromosomes surrounded by a nuclear membrane.
Heritable trans-locationassay	Exposed m ale Drosophila or mice and bred to non exposed females. then bred to detect chromosomal translocatioons.
Sister Chromatid	Mice are exposed to a substance and their bone marrow
Exchange	cells or lymphocytes are microscopically exmined for
assay (SCE)	chromosome breakage. This is indicated by chromatid fragments joining sister chromatids rather than their own.

In vitro tests for chromosomal effects involve the exposure of cell cultures and microscopic examination for chromosome damage. The most commonly used cell lines are Chinese Hamster Ovary (CHO) cells and human lymphocyte cells. The CHO cells are easy to culture, grow rapidly, and have a low chromosome number (22) which makes for easier identification of chromosome damage.

Human lymphocytes are more difficult to culture. They are obtained from healthy human donors with known medical histories. The results of these assays are potentially more relevant to determine effects of xenobiotics that induce mutations in humans.

Two widely used genotoxicity tests measure DNA damage and repair which is not mutagenicity. DNA damage is considered the first step in the process of mutagenesis. The most commonly used test for unscheduled DNA synthesis (UDS) involves exposure of mammalian cells in culture to a test substance. UDS is measured by the uptake of tritium-labeled thymidine into the DNA of the cells. Rat hepatocytes or human fibroblasts are the common mammalian cell lines used.

Another assay to detect DNA damage involves the exposure of repair-deficient E. coli or B. subtilis. DNA damage can not be repaired so the cells die or their growth may be inhibited.

Toxicology and Toxicity Testing

Toxicology is the science of poisons. Poisons are chemical substances that are harmful or "toxic" to living things. All substances are poisonous if ingested in sufficient quantities, including such necessities of life as water and oxygen (Table 3.1). Humans and animals can be exposed to both naturally occurring and man-made chemicals in a variety of ways — by mouth, skin contact, or inhalation. Toxicological tests measure the effects of a limited exposure of an animal to a substance (acute toxicity) as well as repeated, long-term exposure (chronic toxicity). Substances are also tested for more specific endpoints such as cytotoxicity (ability to damage cells), mutagenicity (ability to cause changes in genetic material), carcinogenicity (ability to cause cancer), and teratogenicity (ability to cause birth defects).

Table 3.1: Approximate LD50 for Various Chemical Substances Fed to Humans

Chemical	Equivalent* LD50 for
	160 lb. Human
Sugar (sucrose)	3 Quarts
Alcohol	3 Quarts
Salt (sodium chloride)	1 Quart
2,3-D	½ Cup
Arsenic (aresenic acid)	1 to 2 Teaspoons
Nicotine	½ teaspoon
Dioxin (TCDD)	Speck
Botulinum toxin	Too small to be seen

All of these are aspects of the process by which substances are evaluated for their potential impact on human health and safety. Risk assessment can be divided into two segments. is an estimate of the number of people who will be exposed to a given chemical, together with concentration, duration, and terms of the exposure. identifies hazards, determining which adverse effects will ensue from exposure to a chemical, and provides data estimating the quantitative for the chemical. The likelihood of a human being or animal developing an adverse response after exposure to a chemical varies depending on the route of exposure (skin contact, ingestion, inhalation) as well as the age, sex, genetic makeup, and health status of the individual. Some, but not all, of the latter group may be assessed by toxicity testing.

Other types of tests establish the and properties of chemicals. Toxicokinetic studies trace the absorption, distribution in the body, metabolism, storage, and excretion of chemicals. Toxicodynamic studies chart biological responses that are a consequence of the presence of a chemical in the body. For example, neurological and behavioral tests monitor effects of chemicals on functions, while tests for determine if sunlight can activate a test chemical and enhance skin irritation.

Why Test?

On April 15, 1980, a coalition led by animal activist Henry Spira took out a full-page advertisement in The New York Times, which posed the incendiary question, "How many rabbits has Revlon blinded for beauty's sake?" Revlon did not offer any figures in answer, but within a year they had donated $750,000 to Rockefeller University to research possible alternatives to the Draize test for ocular irritation, the first corporate funding of the embryonic science of in vitro toxicology.

However, in 1921, the Journal of the American Medical Association did quantify the results of the lack of testing for eye irritation, reporting numerous cases of human blindness and disfigurement (and at least one death) resulting from the use of a synthetic aniline dye called Lash-Lure, which was applied by operators in beauty salons to darken eyebrows and eyelashes. Just how many women have been injured by the use of this preparation there is no way of knowing; but the Journal of the American Medical Association has reported at least 17 authentic cases, and there have no doubt been many others, for the firm has settled a number of what it termed 'nuisance' cases for small sums. Still other claims for damages have been paid by the beauty shops or their insurance companies...(Lamb, 1926).

Lash-Lure was not the only cosmetic preparation to cause injury. A hair dye called Inecto Rapid Notox was even more dangerous. Among the specific injuries recorded in 37 cases where a hair dye containing The Criminal Ingredient was used were swelling of the face, eyelids, and larynx, inflammation of the skin with cracking, blistered scalp, impaired eyesight, swelling of the head and limbs, spreading of infection and eruption over entire body, hospitalization with incapacity for long periods, permanent blindness developing.

U.S. Regulatory Law

Although regulation of foods in the United States originated during colonial times, and federal controls over drugs were enacted in 1848 (Green/Bradlaw, 1992), federal responsibility for consumer protection did not become a reality until the passage of the Food, Drug, and Cosmetics Act of 1938. The Delaney Amendments to this law, passed in 1958, require manufacturers to furnish data establishing the noncarcinogenicity of a product prior to marketing and sales.

The Toxic Substances Control Act of 1976 gives the Environmental Protection Agency (EPA) (established in 1970) the power to ban or restrict the manufacture or use of any chemical that it deems hazardous. The Act also authorizes the EPA to require testing of potentially harmful chemical substances already on the market. The Federal Insecticide, Fungicide, and Rodenticide Act requires that all pesticides distributed in the United States be registered.

The Occupational Safety and Health Administration (OSHA), created in 1971, is responsible for regulating safety in the workplace. OSHA uses both epidemiological and animal studies to make regulatory decisions regarding toxins. Finally, the Consumer Products Safety Commission (CPSC) is responsible for maintaining a clearinghouse of information about the hazards associated with the use of consumer products.

Each of these agencies is charged with ensuring the health and safety of the American people and each has its counterparts in other nations. One of the current challenges facing government scientists in many nations is finding a way to balance regulatory responsibilities with the desire to reduce animal testing. A recent fact sheet outlining current federal practices and regulations with respect to eye and skin irritation notes that the CPSC, EPA, and Federal Drug Administration do not require Draize tests for products or chemicals known to be corrosive, and that Draize tests are the final, not the initial, analysis done on products or chemicals that might prove irritating to the eyes or skin.

These descriptions illuminate the era in which tests such as the Draize test for eye and skin irritation were developed, when the toxins arsenic, quinine, resorcin, and mercury were commonly used in various cosmetics and personal care products. Drugs, too, often proved dangerous to consumers. Over 100 Americans died in 1937 when sulfanilamide, an early antibiotic, was mistakenly mixed with the toxic solvent diethylene glycol, and marketed as Elixir of Sulfanilamide. Not until enactment of the Federal Food, Drug, and Cosmetic Act of 1938 was a government agency given the power to regulate such products. Legislation existing prior to that time "prohibited interstate commerce in misbranded and adulterated foods, drinks and drugs" but did not grant authority over cosmetics nor require safety testing for new drugs or provide for tolerance levels for poisonous substances.

Public emergencies resulting from the leakage or spillage of toxic chemicals have also occurred with regularity in the modern era. From Nitro, West Virginia in 1947 to Seveso, Italy in 1976 to Bhopal, India in 1984, human beings have been exposed to chemical toxins with tragic results. Leaving aside for the moment questions of culpability and liability, the fact remains that as long as human beings manufacture and store large quantities of toxic chemicals, the potential for such catastrophe exists and medical staff members will need basic information to treat victims.

Chemical Disasters

On July 10, 1976 an explosion at an Icmesa factory in Seveso, Italy released 1.3 kg of 2,3,7,8-tetrachlorodibenzodioxin, more commonly known as TCDD or dioxin, into the air. Over the next 3 days, vegetation, birds, and animals sickened and died in an area directly downwind from the plant, designated Zone A. Despite this evidence of toxicity, the 735 human residents of Zone A were not evacuated until July 26, 16 days after the accident. Studies later confirmed that the residents of Zone A exhibited the highest levels of dioxin ever found in human serum and that the soil in the zone was heavily contaminated.

Dioxin is a highly toxic synthetic chemical that resists environmental degradation. Since 1961, scientists have known that human beings exposed to dioxin often develop chloracne, a skin disease. The chemical's proven effects on animals are considerably more serious and include cancer, birth defects, and severe liver damage. Workers exposed to dioxin after a March 8, 1949 explosion at a Monsanto plant in Nitro, West Virginia developed skin and eye irritations, headaches, dizziness, and breathing problems in the immediate aftermath of the incident. Epidemiological studies following the Nitro and Seveso incidents have not proven excess cancer, birth defects, or any clearly substantiated toxicoses other than chloracne in these human populations. However, anecdotal accounts of neurological damage, decreased sexual potency in men, and lower birthrates in women persist.

In the early morning hours of December 3, 1984, 200,000 people in Bhopal, India were exposed to the gas methyl isocyanate. The 90-minute exposure resulted in at least 2,500 deaths and countless cases of severe eye and lung damage. Most of the deaths were caused by pulmonary edema (excess accumulation of fluid in the lungs) or its effects. Medical treatment of the victims was handicapped by a lack of information on the effects of methyl isocyanate in human beings; a lack that the accident has, unfortunately, remedied.

Furthermore, normal discharges and emissions may also pose a hazard over the long run. The National Cancer Institute is currently funding a large-scale study of the link between environmental contaminants and breast cancer, based upon the results of a recent study published in the Archives of Environmental Health. This study at Hartford Hospital in Connecticut found that fat samples taken from malignant breast tumors contained more than 50% more PCBs (polychlorinated biphenyls) and DDT (dichlorodiphenyltrichloroethane) than were found in samples taken from women of the same age and weight with noncancerous breast biopsies.

Most of our current understanding about the toxicity of various chemicals comes from animal data, and as Andrew Rowan notes in his critical evaluation of animal research, Of Mice, Models & Men, "there is not doubt that our knowledge of the risks to humans of most chemicals is very inadequate". He and a number of other commentators have pointed out that an increase in animal testing, such as that which occurred after the thalidomide disaster in 1962, will hardly address the problem. "The pressing need at the moment is for more data on mechanisms and the development of a theoretical framework within which rational decision making can have a chance".

The Truth About Thalidomide

Fetal malformations caused by the use of the drug thalidomide constitute one of the most tragic chapters in modern pharmacology. Over the past decade, many individuals and organizations have used this episode to illustrate the inadequacy of animal testing, pointing out that extensive testing in animals did not reveal the teratogenic potential of the drug in human beings. However, as Rowan revealed in 1984, this claim is erroneous. "The fact is that thalidomide was not adequately tested, and after the tragedy, drug registration authorities around the world immediately increased their animal-testing requirements" (Rowan, 1984).

Although the drug was marketed in 1957, reproductive studies on thalidomide in animals were not started until 1961, after the drug's effects on human fetuses had begun to be suspected. Initial studies on rats and mice revealed some reproductive abnormalities, notably reduction in litter size due to resorption of fetuses; however, only when the compound was tested in the New Zealand white rabbit did abnormalities similar to those noticed in human babies occur. Studies on monkeys revealed that they were almost as sensitive as humans to the deformative effects of the drug.

Rowan traces the confusion about thalidomide to the publication of Richard Ryder's book, Victims of Science, noting that Ryder may have been misled by the claims of the Turkish scientist Aygun that he had evidence of the reproductive toxicity of thalidomide in tissue culture. This allegation has never been substantiated. The Insight Team of The Sunday Times of London revealed that neither the German makers of the drug nor the British distributors performed any type of premarket teratogenic testing on animals, although such tests were being performed at that time on other sedative drugs such as Miltown and Librium. If anything, the story of thalidomide exhibits the need for tight regulation and extensive testing of new drugs and chemicals. When the German manufacturer of the drug was asked in 1961, four years after the drug went on the market, "when thalidomide is given to women patients, does it cross the placenta" the company's answer was "not known." The answer was not known because the necessary tests had not been performed.

Knowledge of the manner in which a chemical's structure determines its activity or how it exerts its ultimate toxic effects is not gained through whole-animal tests alone, but also through in vitro studies. In vitro biomedical research attempts to preserve organs, tissues, and cells outside the body. These cell cultures, tissue cultures, or organ cultures can then be used for a number of purposes, including toxicity testing.

The advantages of in vitro systems in toxicity testing are numerous. In vitro tests are usually quicker and less expensive. Experimental conditions can be highly controlled and the results are easily quantified. However, the relative simplicity of nonwhole-animal testing results in limitations as well. Cells or tissues in culture cannot predict the effect of a toxin on a living organism with its complex interaction of nervous, endocrine, immune, and hematopoietic systems. In vitro systems can predict the cellular and molecular effects of a drug or toxin, but only a human or animal can exhibit the complex physiological response of the whole organism, including signs and symptoms of injury.

What Is an Alternative?

As the science of in vitro toxicology has grown over the past few years, some confusion has developed over the exact definition of the word "alternative" as applied to these methodologies. A replacement alternative is one that entirely eliminates the need for whole-animal testing. The limulus assay for bacterial endotoxins, in which the fever-producing potentialities of intravenous therapies are tested using the (extracted) blood of horseshoe crabs rather than whole rabbits, is one such replacement alternative. The use of in vitro systems for pregnancy testing is another (Fig. 3.1).

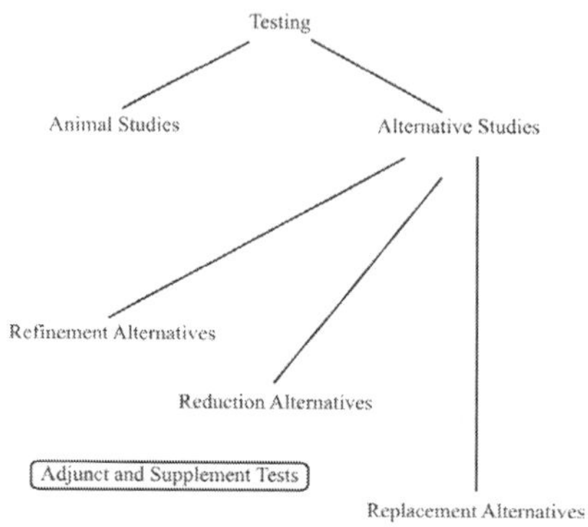

Fig. The three R's of alternatives — replacement, reduction, and refinement.

A reduction alternative substantially decreases the number of whole animals necessary to perform a particular test or group of tests. A number of in vitro assays are now being used as screening tests for the Draize test for ocular irritation, reducing the number of animals required to fully evaluate the potential irritancy of a chemical. Refinement alternatives are those that improve the design and/or efficiency of the test, therefore lessening the distress or discomfort experienced by laboratory animals. Draize tests are no longer performed on substances with a pH of less than 2.0 or more than 11.5 (very acidic and very basic substances known to be severely irritating to the eye). This is a refinement alternative.

At present, in vitro methods are being developed in laboratories around the world. In a comparatively brief period of time, an international consensus has developed regarding the necessity of developing alternatives to the use of animals in toxicity testing. This is partly in response to citizen concern about animal welfare, but it is also an aspect of the evolution of science itself.

European Ban on Animal Testing

On February 12, 1992, the European Parliament voted to amend Cosmetics Directive 76/768 to ban the marketing of cosmetics containing ingredients that have been tested on animals after January 1, 1998. The Council of Ministers agreed to consider extending the 1998 deadline for a period of not less than two years, if validated nonanimal tests have not been developed. With 518 members representing 320 million people, the European Parliament represents the world's largest trading group, the 12-nation European community. The European legislation will affect all companies in that they will be unable to market in Europe products that have been tested on animals, or containing ingredients that have been tested on animals, even if those products have been manufactured outside of Europe. Industrial chemicals are exempt from the ban.

Directive 76/768 was amended after an extensive campaign by national and international animal protectionist groups. A petition containing 2.5 million signatures was presented to the Chairman of the Parliament's Environment committee in June 1992. Rallies and legislative lobbying also contributed to the success of the campaign. At this point, it is not possible to assess the impact of the European legislation on American and international law or industrial practice. However, the amendment of this directive demonstrates that animal protectionism is a potent political force, capable of achieving its goals within large legislative bodies.

In the past 40 years bioscience has endured a change of perspective as profound as that borne by physics when the world was greeted with Einstein's famous equation E=mc2. The shift in scientific perspective, which arose in response to Einstein's theories of relativity and the contiguously developed quantum theory, has been paralleled by that effected by the discoveries in molecular biology. In both cases the eye of science has been drawn inward to the infinitesimal, the nearly invisible processes that create and sustain life. In physics, the stage in which these processes are enacted is the atom; in biology it is the cell, with its complex organization of molecules and information systems (Fig. 3.2).

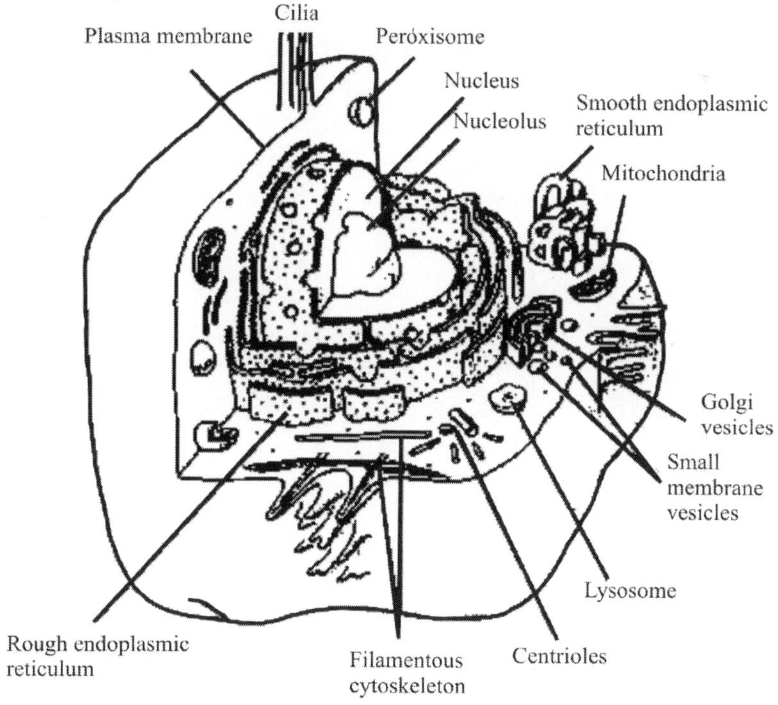

Fig. 3.2: Sketch of a cell with key structures noted, taken from an introductory text in molecular biology.

As a result of this shift in perspective and related discoveries, bioscience is in the midst of the same sort of technological revolution that physical science underwent 100 years ago when the discovery of the laws of electromagnetism and thermodynamics led to the production of electric motors, electric lights, telephones, radios, and most of the other modern conveniences that we take for granted today.

Researchers are just beginning to develop the knowledge base to produce the tools necessary to manipulate biological processes and solve the problems associated with disease. However, the biotechnological enterprise is based not on the macrocosm as was the Industrial Revolution, but on the microcosm, the barely perceptible processes that take place in molecules and cells.

As Thomas Kuhn noted in his seminal treatise on scientific paradigms and their replacement, The Structure of Scientific Revolutions, "we do not see electrons, but rather their tracks, or else bubbles or vapor in a cloud chamber". Similarly, molecular biologists do not see the process of protein synthesis itself but rather, the evidence of protein synthesis. Bioscience, like physics, has turned inward and penetrated deeper and deeper into a nearly invisible world.

Over the past 20 years it has become clear that the process of writing a sentence involves chemical and molecular interactions as complex and delicate as the physical processes that keep the moon from crashing into earth or the stars from falling from the sky. A message is sent from the brain through the arm to the hand. It sounds simple and familiar enough, but consider the thousands of neural and glial cells, synapses, transmitters, and receptors that carry out the simple process; a single exposure to a potent neurotoxic substance is enough to sabotage the process and turn this sentence to aeua;/dnaa/f;a.

So how does a scientist determine if a new pesticide, or hairspray, or household cleaner is going to disrupt the nervous system, causing the consumer to write and speak gibberish, or walk unsteadily, or fall into a coma? In a typical in vivo study, a scientist trained in toxicology will select a group of animals, use half as a control group, and administer the substance to the experimental group while observing their reactions, prior to sacrificing them and examining their tissues. The toxicologist then renders a report to his or her agency (if a government regulatory toxicologist) or to his or her company (if an industrial toxicologist) and a decision regarding the use of the chemical is made.

This process is an aspect of risk assessment, the primary mechanism by which the health of the public is protected. No one wants to eliminate risk assessment. However, since 1981, a number of individuals and organizations have been working very hard to develop alternatives to the traditional whole-animal models used in risk assessment. In most cases they are attempting to discover the cellular and molecular mechanisms by which a substance exerts its toxic effects as well as recording the effects themselves. This sounds very easy but is in actuality incredibly difficult. Researchers in this new field, in vitro toxicology, are working on the cutting edge of knowledge and technique, which encompasses molecular and cellular biology, complex biochemistry, and biophysics.

In this endeavor they have been supported by the three R concept of replacement, reduction, and refinement first articulated in The Principles of Humane Experimental Technique. Written by W.M.S. Russell and Rex Burch of the Universities Federation for Animal Welfare (UFAW) and published in 1959, The Principles of Humane Experimental Technique marks an important scientific attempt to define and, hence, alleviate the distress suffered by laboratory animals. Russell and Burch stated that good science and animal welfare are not incompatible; rather the two go hand in hand. They found that experimental results taken from stressed animals may be misleading or erroneous. Replacement, reduction, and refinement are therefore essential to the creation of a more precise science, as well as a more humane one.

Serendipitously, Russell and Burch's proposal of the three Rs has coincided with the development of molecular biology, as well as with vast improvements in technique for such things as propagating cells in culture and measuring molecular processes. As Rowan notes in Of Mice, Models & Men, "many important discoveries are preceded by the development of one or more techniques of suitable sensitivity and discriminative powers".

Biotechnology and Alternatives

Biotechnology, the use of biological processes to manufacture products, will have a significant impact on the development of alternatives and has already made contributions to the science of in vitro toxicology. Over the past decade, advances in culture methodology, including the developments of defined media, matrices, and effective growth factors have brought cell and tissue culture to new levels of sophistication. Researchers at The Johns Hopkins School of Medicine recently developed two human cortical neuronal cell lines for use in toxicity testing and basic research. Such a feat was almost inconceivable 20 years ago and considered impossible even a decade ago - the stuff of science fiction. Today it is a reality thanks to basic research and biotechnology.

Biotechnology will drive the development of in vitro methodologies in yet another way. As new biochemical entities, such as human growth hormones, are produced they will be tested, just as prior chemical formulations have been tested.

However, in many cases, the only way to evaluate the efficacy and safety of these new compounds will be in cultures of human cells or through epidemiological studies. The nature of the technology will therefore mandate the use of human, not animal, tissues and subjects.

One other factor has contributed to the diminishing use of animals as models in toxicity testing. The animal rights movement and the political pressures brought to bear by animal protectionists have had (and will likely continue to have) noticeable impact on the practice of biomedical research. Ever since the publication of Rachel Carson's Silent Spring in 1962 instigated a societal reevaluation of human behavior and its impact on ecosystems, thoughtful people have recognized that the earth and its resources, whether animal, vegetable, or mineral, should be treated with respect and perhaps even reverence. That perspective may not preclude the use of animals in biomedical research or as food or companions, but it certainly mitigates against their being viewed as "test tubes with legs," as the authors of The Animal Rights Crusade: The Growth of a Moral Protest points out (Jasper and Nelkin, 1992).

At this stage in the alternatives debate, most toxicologists are willing to accept the fact that in vitro and other alternatives are a fruitful avenue for research and that at some point in the future they may replace many whole-animal methods in toxicity testing. Most also insist that until that time in vitro and other alternatives should serve as screens for in vivo (within the living organisms) tests. In addition, it is not possible to predict what scientific information will be produced over the next decade and what its effect on the feasibility and timing of replacement methodologies will be.

Those who argue that in vitro tests can never truly replace whole-animal methods illustrate a concept that Thomas Kuhn termed incommensurability. "The proponents of competing paradigms tend to talk at cross purposes - each paradigm is shown to satisfy more or less the criteria that it dictates for itself and to fall short of a few of those dictated by its opponent" (Kuhn, 1962). Critics of whole-animal methods claim that species variability invalidates the animal model, and proponents of traditional in vivo models insist that in vitro tests cannot predict the systemic effects of a toxic substance in a whole animal or human subject.

The problem is exacerbated by the fact that in vitro toxicology is an infant science, hardly mature enough to satisfy the enormous demands being placed upon it by those who would like whole-animal testing to be abolished this year, next year, or the year after that. Kuhn notes that it takes at least a generation for a cycle of change to complete itself. Until the cycle is complete "there will be a large but never complete overlap between the problems that can be solved by the old and by the new paradigm.

4

AIR POLLUTION AND TOXICOLOGY

Air pollution can be defined as an alteration of air quality that can be characterized by measurements of chemical, biological or physical pollutants in the air. Therefore, air pollution means the undesirable presence of impurities or the abnormal rise in the proportion of some constituents of the atmosphere. It can be classified in 2 sections: visible and invisible air pollution.

Air pollution refers to the release of pollutants into the air—pollutants which are detrimental to human health and the planet as a whole. According to the World Health Organization (WHO), each year air pollution is responsible for nearly seven million deaths around the globe. Nine out of ten human beings currently breathe air that exceeds the WHO's guideline limits for pollutants, with those living in low- and middle-income countries suffering the most. In the United States, the Clean Air Act, established in 1970, authorizes the U.S. Environmental Protection Agency (EPA) to safeguard public health by regulating the emissions of these harmful air pollutants.

THE GLOBAL REACH OF AIR POLLUTION

Because some air pollutants persist in the atmosphere and are carried long distances by winds, air pollution transcends local, regional, and continental boundaries, and it also may have an effect on global climate and weather. For example, acid rain has gained worldwide attention since the 1970s as a regional and even continental problem. Acid rain occurs when sulfur dioxide and nitrogen oxides from the burning of fossil fuels combine with water vapour in the atmosphere, forming sulfuric acid and nitric acid mists. The resulting acidic precipitation is damaging to water, forest, and soil resources. It has caused the disappearance of fish from many lakes in the Adirondack Mountains of North America, the widespread death of forests in mountains of Europe, and damage to tree growth in the United States and Canada. Acid rain can also corrode building materials and be hazardous to human health. These problems are not contained by political boundaries. Emissions from the burning of fossil fuels in the middle sections of the United States and Canada are precipitated as acid rain in the eastern regions of those countries, and acid rain in Norway comes largely from industrial areas in Great Britain and continental Europe. The international scope of the problem has led to the signing of international agreements on the limitation of sulfur and nitrogen oxide emissions.

Another global problem caused by air pollution is the ozone depletion in the stratosphere. At ground level (i.e., in the troposphere), ozone is a pollutant, but at altitudes above 12 km (7 miles) it plays a crucial role in absorbing and thereby blocking ultraviolet radiation (UV) from the Sun before it reaches the ground. Exposure to UV radiation has been linked to skin cancer and other health problems. In 1985 it was discovered that a large "ozone hole," an ozone-depleted region, is present every year between August and November over the continent of Antarctica. The size of this hole is increased by the presence in the atmosphere of chlorofluorocarbons (CFCs); these emanate from aerosol spray cans, refrigerators, industrial solvents, and other sources and are transported to Antarctica by atmospheric circulation. It had already been demonstrated in the mid-1970s that CFCs posed a threat to the global ozonosphere, and in 1978 the use of CFCs as propellants in aerosol cans was banned in the United States. Their use was subsequently restricted in several other countries. In 1987 representatives from more than 45 countries signed the Montreal Protocol, agreeing to place severe limitations on the production of CFCs. The efficacy of this legislation and global effort can be seen in the ozone layer's continued recovery; in 2019 scientists recorded the smallest ozone hole above Antarctica since 1982.

One of the most significant effects of air pollution is on climate change, particularly global warming. As a result of the growing worldwide consumption of fossil fuels, carbon dioxide levels in the atmosphere have increased steadily since 1900, and the rate of increase is accelerating. It has been estimated that if carbon dioxide levels are not reduced, average global air temperatures may rise another 4 °C (7.2 °F) by the end of the 21st century. Such a warming trend might cause melting of the polar ice caps, rising of the sea level, and flooding of the coastal areas of the world. Changes in precipitation patterns caused by global warming might have adverse effects on agriculture and forest ecosystems, and higher temperatures and humidity might increase the incidence of disease in humans and animals in some parts of the world. Implementation of international agreements on reducing greenhouse gases are required to protect global air quality and to mitigate the effects of global warming.

Indoor air pollution

Health risks related to indoor air pollution have become an issue of concern because people generally spend most of their time indoors at home and at work. The problem has been exacerbated by well-meaning efforts to lower air-exchange rates in buildings in order to conserve energy; these efforts unfortunately allow contaminants to accumulate indoors. Indoor air pollutants include various combustion products from stoves, kerosene space heaters, and fireplaces, as well as volatile organic compounds (VOCs) from household products (e.g., paints, cleaning agents, and pesticides). Formaldehyde off-gassing from building products (especially particleboard and plywood) and from dry-cleaned textiles can accumulate in indoor air. Bacteria, viruses, molds, animal dander, dust mites, and pollen are biological contaminants that can cause disease and other health problems, especially if they build up in and are spread by central heating or cooling systems. Environmental tobacco smoke, also called secondhand smoke, is an indoor air pollutant in many homes, despite widespread knowledge about the harmful effects of smoking. Secondhand smoke contains many carcinogenic compounds as well as strong irritants. In some geographic regions, naturally occurring radon, a radioactive gas, can seep from the ground into buildings and accumulate to harmful levels. Exposure to all indoor air pollutants can be reduced by appropriate building construction and maintenance methods, limitations on pollutant sources, and provision of adequate ventilation.

Types of Air Pollutants

In another lesson, we learned about the Clean Air Act of 1970, which set limits on emissions and standards for air quality, provided funding for pollution control research and made it possible for citizens to sue those who violate the standards. The standards set forth in this legislation by the Environmental Protection Agency, or EPA, identified six pollutants as those that posed the greatest threat to human health. Let's look at each one.

Carbon monoxide is both odorless and colorless and it's produced through combustion. Most carbon monoxide emissions come from vehicles, about 62%, but they also come from forest fires, lawn equipment and the burning of industrial waste. This pollutant is dangerous to human health because, even in very small concentrations, it can prevent oxygen from being delivered through your body to major organs. At higher levels, it can cause death.

Sulfur dioxide is also a colorless gas, and it's highly reactive. About 70% of sulfur dioxide emissions come from combustion at industrial power plants, like those that generate electricity from coal. Sulfur in the coal reacts with atmospheric oxygen in the air and forms the compound sulfur dioxide. Once in the air it may react further to create sulfur acid, which can fall back to Earth as acid rain. Sulfur dioxide can also cause respiratory illnesses when it's breathed in.

Nitrogen oxides are a family of gases that are also very reactive, but you can definitely see and smell these! Both nitric oxide and nitrogen dioxide are in this family, and these are the gases that contribute to the smog you see over large cities, like Los Angeles. More than half of these pollutants come from vehicle combustion emissions, with a large amount also coming from industrial combustion. Nitrogen oxides also cause respiratory illnesses.

Ozone is a tricky one because when it's in the stratosphere layer of the atmosphere it protects us from harmful UV radiation. However, when ozone is lower in the atmosphere, it becomes dangerous to our health because it causes respiratory illnesses and damages ecosystems. Ozone is a secondary pollutant because it reacts from a combination of sunlight, heat, nitrogen oxides and volatile carbon containing compounds.

What Are the Six Common Air Pollutants

The Clean Air Act requires EPA to set National Ambient Air Quality Standards for six common air pollutants. These commonly found air pollutants (also known as "criteria pollutants") are found all over the United States. They are particle pollution (often referred to as particulate matter), ground-level ozone, carbon monoxide, sulfur oxides, nitrogen oxides, and lead. These pollutants can harm your health and the environment, and cause property damage. Of the six pollutants, particle pollution and ground-level ozone are the most widespread health threats. EPA calls these pollutants "criteria" air pollutants because it regulates them by developing human health-based and/or environmentally-based criteria (science-based guidelines) for setting permissible levels. The set of limits based on human health is called primary standards. Another set of limits intended to prevent environmental and property damage is called secondary standards.

Click on one of the pollutants below for information on sources of the pollutant, why the pollutant is of concern, health and environmental effects, efforts underway to help reduce the pollutant, and other helpful resources.

- Ozone
- Particulate Matter

- Carbon Monoxide
- Nitrogen Oxides
- Sulfur Dioxide
- Lead

Air Pollution Trends

For each of these pollutants, EPA tracks two kinds of air pollution trends: air concentrations based on actual measurements of pollutant concentrations in the ambient (outside) air at selected monitoring sites throughout the country, and emissions based on engineering estimates of the total tons of pollutants released into the air each year. Despite the progress made in the last 30 years, millions of people live in counties with monitor data showing unhealthy air for one or more of the six common air pollutants. For EPA's most recent evaluation of air pollution trends for these six pollutants, click on the following:

- Latest Findings on National Air Quality: Status and Trends

Health Effects Information

Exposure to these pollutants is associated with numerous effects on human health, including increased respiratory symptoms, hospitalization for heart or lung diseases, and even premature death. Try these helpful resources:

- Air Quality Index (AQI)
- Ozone - Good Up High, Bad Nearby
- Ozone and Your Health
- Particle Pollution and Your Health
- Air Quality Guide for Ozone
- Air Quality Guide for Particle Pollution
- Smog - Who Does it Hurt?

STATE IMPLEMENTATION PLAN STATUS AND INFORMATION

EPA must designate areas as meeting (attainment) or not meeting (nonattainment) the standard. The Clean Air Act (CAA) requires states to develop a general plan to attain and maintain the NAAQS in all areas of the country and a specific plan to attain the standards for each area designated nonattainment for a NAAQS. These plans, known as State Implementation Plans or SIPs, are developed by state and local air quality management agencies and submitted to EPA for approval. Detailed information about state SIP elements and their status can be found on the State Implementation Plan Status and Information page.

Sources of Air Pollution

When you think of air pollution, you likely think of human activities, such as coal-burning plants and emissions from cars, planes and boats. But, air pollution also comes from natural sources. Volcanoes are a great example of a natural pollution source; they produce large amounts of particles and gases, and the eruptions send them up into the atmosphere. Forest fires are a natural process that also produces large amounts of potentially harmful gas and particulate matter.

No matter the source, there are two ways pollution can enter the air. Point source pollution is when the air pollutants come from a single source of origin, such as smokestacks at a single factory. Non-point source pollution is when the air pollutants come from many sources, such as all of the cars in the U.S.

Just like not all sources of pollution are the same, pollutants also vary in their effects. Primary pollutants are those that cause direct harm or that can react to form harmful substances in the atmosphere. Secondary pollutants are those harmful substances that are created from the reactions between primary pollutants and the components of the atmosphere.

Air pollution can result from both human and natural actions. Natural events that pollute the air include forest fires, volcanic eruptions, wind erosion, pollen dispersal, evaporation of organic compounds and natural radioactivity. Pollution from natural occurrences are not very often.

Human activities that result in air pollution include

1. Emissions from industries and manufacturing activities

Consider a typical manufacturing plant: You will notice that there are long tubes (called chimneys) erected high into the air, with lots of smoke and fumes coming out of it. Waste incinerators, manufacturing industries and power plants emit high levels of carbon monoxide, organic compounds, and chemicals into the air. This happens almost everywhere that people live. Petroleum refineries also release lots of hydrocarbons into the air.

2. Burning Fossil Fuels

After the industrial age, transportation has become a key part of our lives. Cars and heavy duty trucks, trains, shipping vessels and airplanes all burn lots of fossil fuels to work. Emissions from automobile engines contain both primary and secondary pollutants. This is a major cause of pollution, and one that is very difficult to manage. This is because humans rely heavily on vehicles and engines for transporting people, good and services.

Fumes from car exhauts contain dangerous gases such as carbon monoxide, oxides of nitrogen, hydrocarbons and particulates. On their own, they cause great harm to people who breath them. Additionally, they react with environmental gases to create further toxic gases. Click here to see the effects

3. Household and Farming Chemicals

Crop dusting, fumigating homes, household cleaning products or painting supplies, over the counter insect/pest killers, fertilizer dust emit harmful chemicals into the air and cause pollution. In many case, when we use these chemicals at home or offices with no or little ventilation, we may fall ill if we breathe them.

What are the common air pollutants around?

Carbon Monoxide (CO)

Carbon Monoxide (CO)

Carbon Monoxide (CO)

- Fuel combustion from vehicles and engines.
- Reduces the amount of oxygen reaching the body's organs and tissues; aggravates heart disease, resulting in chest pain and other symptoms.

Ground-level Ozone (O_3)

- Secondary pollutant formed by chemical reaction of volatile organic compounds (VOCs) and

NOx in the presence of sunlight.

- Decreases lung function and causes respiratory symptoms, such as coughing and shortness of breath, and also makes asthma and other lung diseases get worse. More on Ground Level Ozone Here

LEAD (PB)

- Smelters (metal refineries) and other metal industries; combustion of leaded gasoline in piston engine aircraft; waste incinerators (waste burners), and battery manufacturing.

Nitrogen Dioxide (NO_2)

- Fuel combustion (electric utilities, big industrial boilers, vehicles) and wood burning.
- Worsens lung diseases leading to respiratory symptoms, increased susceptibility to respiratory infection.

Particulate Matter (PM)

- This is formed through chemical reactions, fuel combustion (e.g., burning coal, wood, diesel), industrial processes, farming (plowing, field burning), and unpaved roads or during road constructions.
- Short-term exposures can worsen heart or lung diseases and cause respiratory problems. Long-term exposures can cause heart or lung disease and sometimes premature deaths.

Sulfur Dioxide (SO_2)

- SO_2 comes from fuel combustion (especially high-sulfur coal); electric utilities and industrial processes as well as natural occurances like volcanoes.
- Aggravates asthma and makes breathing difficult. It also contributes to particle formation with associated health effects.

WHAT ARE THE EFFECTS OF AIR POLLUTION?

Below are a few key effects of air pollution. The last page on FactSheet has some specific air pollution incidents that are worth checking out.

Acidification

Chemical reactions involving air pollutants can create acidic compounds which can cause harm to vegetation and buildings. Sometimes, when an air pollutant, such as sulfuric acid combines with the water droplets that make up clouds, the water droplets become acidic, forming acid rain. When acid rain falls over an area, it can kill trees and harm animals, fish, and other wildlife.

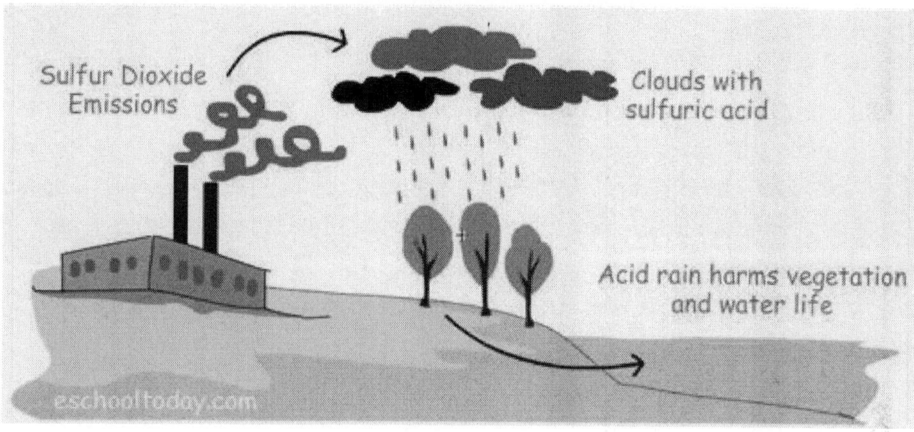

Acid rain destroys the leaves of plants

When acid rain infiltrates into soils, it changes the chemistry of the soil making it unfit for many living things that rely on soil as a habitat or for nutrition. Acid rain also changes the chemistry of the lakes and streams that the rainwater flows into, harming fish and other aquatic life.

Eutrophication

Rain can carry and deposit the Nitrogen in some pollutants on rivers and soils. This will adversely affect the nutrients in the soil and water bodies. This can result in algae growth in lakes and water bodies, and make conditions for other living organism harmful.

Ground-level ozone

Chemical reactions involving air pollutants create a poisonous gas ozone (O3). Gas Ozone can affect people's health and can damage vegetation types and some animal life too.

Particulate matter

Air pollutants can be in the form of particulate matter which can be very harmful to our health. The level of effect usually depends on the length of time of exposure, as well the kind and concentration of chemicals and particles exposed to. Short-term effects include irritation to the eyes, nose and throat, and upper respiratory infections such as bronchitis and pneumonia. Others include headaches, nausea, and allergic reactions. Short-term air pollution can aggravate the medical conditions of individuals with asthma and emphysema. Long-term health effects can include chronic respiratory disease, lung cancer, heart disease, and even damage to the brain, nerves, liver, or kidneys. Continual exposure to air pollution affects the lungs of growing children and may aggravate or complicate medical conditions in the elderly.

Air pollution is the introduction of particulates, biological molecules, or other harmful materials into Earth's atmosphere, causing disease, death to humans, damage to other living organisms such as food crops, or the natural or built environment. Air pollution may come from anthropogenic or natural sources. The atmosphere is a complex natural gaseous system that is essential to support life on planet Earth. Stratospheric ozone depletion due to air pollution has been recognized as a threat to human health as well as to the Earth's ecosystems. Indoor air pollution and urban air quality are listed as two of the world's worst toxic pollution problems in the 2008 Blacksmith Institute World's Worst Polluted Places report. According to the 2014 WHO report, air pollution in 2012 caused the deaths of around 7 million people worldwide.

EFFECTS OF AIR POLLUTION

The effects of air pollution on the human body vary depending on the type of pollutant and the length and level of exposure—as well as other factors, including a person's individual health risks and the cumulative impacts of multiple pollutants or stressors.

Smog and soot

These are the two most prevalent types of air pollution. Smog (sometimes referred to as ground-level ozone) occurs when emissions from combusting fossil fuels react with sunlight. Soot (also known as particulate matter) is made up of tiny particles of chemicals, soil, smoke, dust, or allergens—in the form of either gas or solids—that are carried in the air. The sources of smog and soot are similar. "Both come from cars and trucks, factories, power plants, incinerators, engines, generally anything that combusts fossil fuels such as coal, gas, or natural gas," Walke says.

Smog can irritate the eyes and throat and also damage the lungs, especially those of children, senior citizens, and people who work or exercise outdoors. It's even worse for people who have asthma or allergies: these extra pollutants can intensify their symptoms and trigger asthma attacks. The tiniest airborne particles in soot, whether gaseous or solid, are especially dangerous because they can penetrate the lungs and bloodstream and worsen bronchitis, lead to heart attacks, and even hasten death. In 2020 a report from Harvard's T. H. Chan School of Public Health showed COVID-19 mortality rates in areas with more soot pollution were higher than in areas with even slightly less, showing a correlation between the virus's deadliness and long-term exposure to fine particulate matter and illuminating an environmental justice issue.

Because highways and polluting facilities have historically been sited in or next to low-income neighborhoods and communities of color, the negative effects of this pollution have been disproportionately experienced by the people who live in these communities. In 2019 the Union of Concerned Scientists found that soot exposure was 34 percent higher for Asian Americans, on average, than for other Americans. For Black people, the exposure rate was 24 percent higher; for Latinos, 23 percent higher.

Hazardous air pollutants

A number of air pollutants pose severe health risks and can sometimes be fatal even in small amounts. Almost 200 of them are regulated by law; some of the most common are mercury, lead, dioxins, and benzene. "These are also most often emitted during gas or coal combustion, incinerating, or—in the case of benzene—found in gasoline," Walke says. Benzene, classified as a carcinogen by the EPA, can cause eye, skin, and lung irritation in the short term and blood disorders in the long term. Dioxins, more typically found in food but also present in small amounts in the air, can affect the liver in the short term and harm the immune, nervous, and endocrine systems as well as reproductive functions. Mercury attacks the central nervous system. In large amounts, lead can damage children's brains and kidneys, and even minimal exposure can affect children's IQ and ability to learn.

Another category of toxic compounds, polycyclic aromatic hydrocarbons (PAHs), are by-products of traffic exhaust and wildfire smoke. In large amounts they have been linked to eye and lung irritation, blood and liver issues, and even cancer. In one study, the children of mothers exposed to PAHs during pregnancy showed slower brain-processing speeds and more pronounced symptoms of ADHD.

Greenhouse gases

By trapping the earth's heat in the atmosphere, greenhouse gases lead to warmer temperatures, which in turn lead to the hallmarks of climate change: rising sea levels, more extreme weather, heat-related deaths, and the increased transmission of infectious diseases. In 2018 carbon dioxide accounted for 81 percent of the country's total greenhouse gas emissions, and methane made up 10 percent. "Carbon dioxide comes from combusting fossil fuels, and methane comes from natural and industrial sources, including large amounts that are released during oil and gas drilling," Walke says. "We emit far larger amounts of carbon dioxide, but methane is significantly more potent, so it's also very destructive." Another class of greenhouse gases, hydrofluorocarbons (HFCs), are thousands of times more powerful than carbon dioxide in their ability to trap heat. In October 2016 more than 140 countries reached an agreement to reduce the use of these chemicals—which are found in air conditioners and refrigerators—and develop greener alternatives over time. Though President Trump was unwilling to sign on to this agreement, a bipartisan group of senators overrode his objections in 2020 and set the United States on track to slash HFCs by 85 percent by 2035. According to David Doniger, senior strategic director of NRDC's Climate and Clean Energy program, "the agreed-to HFC phasedown will avoid the equivalent of more than 80 billion tons of carbon dioxide over the next 35 years."

Pollen and mold

Mold and allergens from trees, weeds, and grass are also carried in the air, are exacerbated by climate change, and can be hazardous to health. Though they aren't regulated and are less directly connected to human actions, they can be considered a form of air pollution. "When homes, schools, or businesses get water damage, mold can grow and can produce allergenic airborne pollutants," Knowlton says. "Mold exposure can precipitate asthma attacks or an allergic response, and some molds can even produce toxins that would be dangerous for anyone to inhale."

Pollen allergies are worsening because of climate change. "Lab and field studies are showing that pollen-producing plants—especially ragweed—grow larger and produce more pollen when you increase the amount of carbon dioxide that they grow in," Knowlton says. "Climate change also extends the pollen production season, and some studies are beginning to suggest that ragweed pollen itself might be becoming a more potent allergen." If so, more people will suffer runny noses, fevers, itchy eyes, and other symptoms.

Air pollution is now the world's fourth-largest risk factor for early death. According to the most recent State of Global Air report—which summarizes the latest scientific understanding of air pollution around the world—4.5 million deaths were linked to outdoor air pollution exposures in 2019, and another 2.2 million deaths were caused by indoor air pollution. "Despite improvements in reducing global average mortality rates from air pollution, the world's most populous countries, India and China, continue to bear the highest burdens of disease," says Vijay Lamaye, staff scientist at the NRDC Science Center. "This report is a sobering reminder that the climate crisis threatens to worsen air pollution problems significantly if we fail to act to cut carbon pollution."

Pollutants

An air pollutant is a substance in the air that can have adverse effects on humans and the ecosystem. The substance can be solid particles, liquid droplets, or gases. A pollutant can be of natural origin or man-made. Pollutants are classified as primary or secondary. Primary pollutants are usually produced from a process, such as ash from a volcanic eruption. Other examples include carbon monoxide gas from motor vehicle exhaust, or the sulfur dioxide released from factories. Secondary pollutants are not emitted directly. Rather, they form in the air when primary pollutants react or interact. Ground level ozone is a prominent example of a secondary pollutant. Some pollutants may be both primary and secondary: they are both emitted directly and formed from other primary pollutants.

Major primary pollutants produced by human activity include:

- Sulfur oxides (SOx) - particularly sulfur dioxide, a chemical compound with the formula SO_2. SO_2 is produced by volcanoes and in various industrial processes. Coal and petroleum often contain sulfur compounds, and their combustion generates sulfur dioxide. Further oxidation of SO2, usually in the presence of a catalyst such as NO2, forms H2SO4, and thus acid rain. This is one of the causes for concern over the environmental impact of the use of these fuels as power sources.
- Nitrogen oxides (NOx) - Nitrogen oxides, particularly nitrogen dioxide, are expelled from high temperature combustion, and are also produced during thunderstorms by electric discharge. They can be seen as a brown haze dome above or a plume downwind of cities. Nitrogen dioxide is a chemical compound with the formula NO_2. It is one of several nitrogen oxides. One of the most prominent air pollutants, this reddish-brown toxic gas has a characteristic sharp, biting odor.
- Carbon monoxide (CO) - CO is a colorless, odorless, toxic yet non-irritating gas. It is a product by incomplete combustion of fuel such as natural gas, coal or wood. Vehicular exhaust is a major source of carbon monoxide.

- Volatile organic compounds (VOC) - VOCs are a well-known outdoor air pollutant. They are categorized as either methane (CH4) or non-methane (NMVOCs). Methane is an extremely efficient greenhouse gas which contributes to enhanced global warming. Other hydrocarbon VOCs are also significant greenhouse gases because of their role in creating ozone and prolonging the life of methane in the atmosphere. This effect varies depending on local air quality. The aromatic NMVOCs benzene, toluene and xylene are suspected carcinogens and may lead to leukemia with prolonged exposure. 1,3-butadiene is another dangerous compound often associated with industrial use.
- Particulates, alternatively referred to as particulate matter (PM), atmospheric particulate matter, or fine particles, are tiny particles of solid or liquid suspended in a gas. In contrast, aerosol refers to combined particles and gas. Some particulates occur naturally, originating from volcanoes, dust storms, forest and grassland fires, living vegetation, and sea spray. Human activities, such as the burning of fossil fuels in vehicles, power plants and various industrial processes also generate significant amounts of aerosols. Averaged worldwide, anthropogenic aerosols—those made by human activities—currently account for approximately 10 percent of our atmosphere. Increased levels of fine particles in the air are linked to health hazards such as heart disease, altered lung function and lung cancer.
- Persistent free radicals connected to airborne fine particles are linked to cardiopulmonary disease.
- Toxic metals, such as lead and mercury, especially their compounds.
- Chlorofluorocarbons (CFCs) - harmful to the ozone layer; emitted from products are currently banned from use. These are gases which are released from air conditioners, refrigerators, aerosol sprays, etc. CFC's on being released into the air rises to stratosphere. Here they come in contact with other gases and damage the ozone layer. This allows harmful ultraviolet rays to reach the earth's surface. This can lead to skin cancer, disease to eye and can even cause damage to plants.
- Ammonia (NH_3) - emitted from agricultural processes. Ammonia is a compound with the formula NH3. It is normally encountered as a gas with a characteristic pungent odor. Ammonia contributes significantly to the nutritional needs of terrestrial organisms by serving as a precursor to foodstuffs and fertilizers. Ammonia, either directly or indirectly, is also a building block for the synthesis of many pharmaceuticals. Although in wide

use, ammonia is both caustic and hazardous. In the atmosphere, ammonia reacts with oxides of nitrogen and sulfur to form secondary particles.

- Odours — such as from garbage, sewage, and industrial processes
- Radioactive pollutants - produced by nuclear explosions, nuclear events, war explosives, and natural processes such as the radioactive decay of radon.

Secondary Pollutants Include

- Particulates created from gaseous primary pollutants and compounds in photochemical smog. Smog is a kind of air pollution. Classic smog results from large amounts of coal burning in an area caused by a mixture of smoke and sulfur dioxide. Modern smog does not usually come from coal but from vehicular and industrial emissions that are acted on in the atmosphere by ultraviolet light from the sun to form secondary pollutants that also combine with the primary emissions to form photochemical smog.
- Ground level ozone (O_3) formed from NOx and VOCs. Ozone (O_3) is a key constituent of the troposphere. It is also an important constituent of certain regions of the stratosphere commonly known as the Ozone layer. Photochemical and chemical reactions involving it drive many of the chemical processes that occur in the atmosphere by day and by night. At abnormally high concentrations brought about by human activities (largely the combustion of fossil fuel), it is a pollutant, and a constituent of smog.
- Peroxyacetyl nitrate (PAN) - similarly formed from NOx and VOCs.

Minor Air Pollutants Include

- A large number of minor hazardous air pollutants. Some of these are regulated in USA under the Clean Air Act and in Europe under the Air Framework Directive
- A variety of persistent organic pollutants, which can attach to particulates

Persistent organic pollutants (POPs) are organic compounds that are resistant to environmental degradation through chemical, biological, and photolytic processes. Because of this, they have been observed to persist in the environment, to be capable of long-range transport, bioaccumulate in human and animal tissue, biomagnify in food chains, and to have potentially significant impacts on human health and the environment.

Sources

There are various locations, activities or factors which are responsible for releasing pollutants into the atmosphere. These sources can be classified into two major categories.

Anthropogenic (man-made) Sources:

These are mostly related to the burning of multiple types of fuel.

- Stationary sources include smoke stacks of power plants, manufacturing facilities (factories) and waste incinerators, as well as furnaces and other types of fuel-burning heating devices. In developing and poor countries, traditional biomass burning is the major source of air pollutants; traditional biomass includes wood, crop waste and dung.
- Mobile sources include motor vehicles, marine vessels, and aircraft.
- Controlled burn practices in agriculture and forest management. Controlled or prescribed burning is a technique sometimes used in forest management, farming, prairie restoration or greenhouse gas abatement. Fire is a natural part of both forest and grassland ecology and controlled fire can be a tool for foresters. Controlled burning stimulates the germination of some desirable forest trees, thus renewing the forest.
- Fumes from paint, hair spray, varnish, aerosol sprays and other solvents
- Waste deposition in landfills, which generate methane. Methane is highly flammable and may form explosive mixtures with air. Methane is also an asphyxiant and may displace oxygen in an enclosed space. Asphyxia or suffocation may result if the oxygen concentration is reduced to below 19.5% by displacement.
- Military resources, such as nuclear weapons, toxic gases, germ warfare and rocketry

Natural Sources:

- Dust from natural sources, usually large areas of land with little or no vegetation
- Methane, emitted by the digestion of food by animals, for example cattle
- Radon gas from radioactive decay within the Earth's crust. Radon is a colorless, odorless, naturally occurring, radioactive noble gas that is formed from the decay of radium. It is considered to be a health hazard. Radon gas from natural sources can accumulate in buildings, especially

in confined areas such as the basement and it is the second most frequent cause of lung cancer, after cigarette smoking.

- Smoke and carbon monoxide from wildfires
- Vegetation, in some regions, emits environmentally significant amounts of Volatile organic compounds (VOCs) on warmer days. These VOCs react with primary anthropogenic pollutants—specifically, NOx, SO2, and anthropogenic organic carbon compounds — to produce a seasonal haze of secondary pollutants. Black gum, poplar, oak and willow are some examples of vegetation that can produce abundant VOCs. The VOC production from these species result in ozone levels up to eight times higher than the low-impact tree species.
- Volcanic activity, which produces sulfur, chlorine, and ash particulates

EMISSION FACTORS

Air pollutant emission factors are representative values that people attempt to relate the quantity of a pollutant released to the ambient air with an activity associated with the release of that pollutant. These factors are usually expressed as the weight of pollutant divided by a unit weight, volume, distance, or duration of the activity emitting the pollutant (e.g., kilograms of particulate emitted per tonne of coal burned). Such factors facilitate estimation of emissions from various sources of air pollution. In most cases, these factors are simply averages of all available data of acceptable quality, and are generally assumed to be representative of long-term averages.

There are 12 compounds in the list of Persistent organic pollutants. Dioxins and furans are two of them and intentionally created by combustion of organics, like open burning of plastics. These compounds are also endocrine disruptors and can mutate the human genes.

The United States Environmental Protection Agency has published a compilation of air pollutant emission factors for a multitude of industrial sources. The United Kingdom, Australia, Canada and many other countries have published similar compilations, as well as the European Environment Agency.

Air Pollution Exposure

Air pollution risk is a function of the hazard of the pollutant and the exposure to that pollutant. Air pollution exposure can be expressed for an individual, for certain groups (e.g. neighborhoods or children living in a county), or for entire populations. For example, one may want to calculate the exposure to a hazardous air pollutant for a geographic area, which includes the various microenvironments and age groups. This can be calculated as an inhalation exposure. This would account for daily exposure in various settings (e.g. different indoor micro-environments and outdoor locations). The exposure needs to include different age and other demographic groups, especially infants, children, pregnant women and other sensitive sub-populations. The exposure to an air pollutant must integrate the concentrations of the air pollutant with respect to the time spent in each setting and the respective inhalation rates for each subgroup for each specific time that the subgroup is in the setting and engaged in particular activities (playing, cooking, reading, working, etc.). For example, a small child's inhalation rate will be less than that of an adult. A child engaged in vigorous exercise will have a higher respiration rate than the same child in a sedentary activity. The daily exposure, then, needs to reflect the time spent in each micro-environmental setting and the type of activities in these settings. The air pollutant concentration in each microactivity/micro-environmental setting is summed to indicate the exposure.

Indoor Air Quality (IAQ)

A lack of ventilation indoors concentrates air pollution where people often spend the majority of their time. Radon (Rn) gas, a carcinogen, is exuded from the Earth in certain locations and trapped inside houses. Building materials including carpeting and plywood emit formaldehyde (H2CO) gas. Paint and solvents give off volatile organic compounds (VOCs) as they dry. Lead paint can degenerate into dust and be inhaled. Intentional air pollution is introduced with the use of air fresheners, incense, and other scented items. Controlled wood fires in stoves and fireplaces can add significant amounts of smoke particulates into the air, inside and out. Indoor pollution fatalities may be caused by using pesticides and other chemical sprays indoors without proper ventilation.

Carbon monoxide poisoning and fatalities are often caused by faulty vents and chimneys, or by the burning of charcoal indoors or in a confined space, such as a tent. Chronic carbon monoxide poisoning can result even from poorly-adjusted pilot lights. Traps are built into all domestic plumbing to keep sewer gas and hydrogen sulfide, out of interiors. Clothing emits tetrachloroethylene, or other dry cleaning fluids, for days after dry cleaning. Though its use has now been banned in many countries, the extensive use of asbestos in industrial and domestic environments in the past has left a potentially very dangerous material in many localities. Asbestosis is a chronic inflammatory medical condition affecting the tissue of the lungs. It occurs after long-term, heavy exposure to asbestos from asbestos-containing materials in structures. Sufferers have severe dyspnea (shortness of breath) and are at an increased risk regarding several different types of lung cancer. As clear explanations are not always stressed in non-technical literature, care should be taken to distinguish between several forms of relevant diseases. According to the World Health Organisation (WHO), these may defined as; asbestosis, lung cancer, and Peritoneal Mesothelioma (generally a very rare form of cancer, when more widespread it is almost always associated with prolonged exposure to asbestos).

Biological sources of air pollution are also found indoors, as gases and airborne particulates. Pets produce dander, people produce dust from minute skin flakes and decomposed hair, dust mites in bedding, carpeting and furniture produce enzymes and micrometre-sized fecal droppings, inhabitants emit methane, mold forms on walls and generates mycotoxins and spores, air conditioning systems can incubate Legionnaires' disease and mold, and houseplants, soil and surrounding gardens can produce pollen, dust, and mold. Indoors, the lack of air circulation allows these airborne pollutants to accumulate more than they would otherwise occur in nature.

Health Effects

Air pollution is a significant risk factor for a number of health conditions including respiratory infections, heart disease, COPD, stroke and lung cancer. The health effects caused by air pollution may include difficulty in breathing, wheezing, coughing, asthma and worsening of existing respiratory and cardiac conditions. These effects can result in increased medication use, increased doctor or emergency room visits, more hospital admissions and premature death. The human health effects of poor air quality are far reaching, but principally affect the body's respiratory system and the cardiovascular system. Individual reactions to air pollutants depend on the type of pollutant a person is exposed to, the degree of exposure, and the individual's health status and genetics. The most common sources of air pollution include particulates, ozone, nitrogen dioxide, and sulfur dioxide. Children aged less than five years that live in developing countries are the most vulnerable population in terms of total deaths attributable to indoor and outdoor air pollution.

Mortality

It is estimated that some 7 million premature deaths may be attributed to air pollution. India has the highest death rate due to air pollution. India also has more deaths from asthma than any other nation according to the World Health Organisation. In December 2013 air pollution was estimated to kill 500,000 people in China each year. There is a correlation between pneumonia-related deaths and air pollution from motor vehicles.

Air pollution is estimated to reduce life expectancy by almost nine months across the European Union. Causes of deaths include strokes, heart disease, COPD, lung cancer, and lung infections.

The US EPA estimates that a proposed set of changes in diesel engine technology (Tier 2) could result in 12,000 fewer premature mortalities, 15,000 fewer heart attacks, 6,000 fewer emergency room visits by children with asthma, and 8,900 fewer respiratory-related hospital admissions each year in the United States. The US EPA estimates allowing a ground-level ozone concentration of 65 parts per billion, would avert 1,700 to 5,100 premature deaths nationwide in 2020 compared with the current 75-ppb standard. The agency projects the stricter standard would also prevent an additional 26,000 cases of aggravated asthma, and more than a million cases of missed work or school.

A new economic study of the health impacts and associated costs of air pollution in the Los Angeles Basin and San Joaquin Valley of Southern California shows that more than 3,800 people die prematurely (approximately 14 years earlier than normal) each year because air pollution levels violate federal standards. The number of annual premature deaths is considerably higher than the fatalities related to auto collisions in the same area, which average fewer than 2,000 per year.

Diesel exhaust (DE) is a major contributor to combustion-derived particulate matter air pollution. In several human experimental studies, using a well-validated exposure chamber setup, DE has been linked to acute vascular dysfunction and increased thrombus formation. This serves as a plausible mechanistic link between the previously described association between particulates air pollution and increased cardiovascular morbidity and mortality.

Cardiovascular Disease

A 2007 review of evidence found ambient air pollution exposure is a risk factor correlating with increased total mortality from cardiovascular events (range: 12% to 14% per 10 microg/m3 increase). Air pollution is also emerging as a risk factor for stroke, particularly in developing countries where pollutant levels are highest. A 2007 study found that in women, air pollution is associated not with hemorrhagic but with ischemic stroke. Air pollution was also found to be associated with increased incidence and mortality from coronary stroke in a cohort study in 2011. Associations are believed to be causal and effects may be mediated by vasoconstriction, low-grade inflammation and atherosclerosis Other mechanisms such as autonomic nervous system imbalance have also been suggested.

Lung Disease

Chronic obstructive pulmonary disease (COPD) includes diseases such as chronic bronchitis and emphysema.

Research has demonstrated increased risk of developing asthma and COPD from increased exposure to traffic-related air pollution. Additionally, air pollution has been associated with increased hospitalization and mortality from asthma and COPD.

A study conducted in 1960-1961 in the wake of the Great Smog of 1952 compared 293 London residents with 477 residents of Gloucester, Peterborough, and Norwich, three towns with low reported death rates from chronic bronchitis. All subjects were male postal truck drivers aged 40 to 59. Compared to the subjects from the outlying towns, the London subjects exhibited more severe respiratory symptoms (including cough, phlegm, and dyspnea), reduced lung function (FEV1 and peak flow rate), and increased sputum production and purulence. The differences were more pronounced for subjects aged 50 to 59. The study controlled for age and smoking habits, so concluded that air pollution was the most likely cause of the observed differences.

It is believed that much like cystic fibrosis, by living in a more urban environment serious health hazards become more apparent. Studies have shown that in urban areas patients suffer mucus hypersecretion, lower levels of lung function, and more self-diagnosis of chronic bronchitis and emphysema.

Cancer

A review of evidence regarding whether ambient air pollution exposure is a risk factor for cancer in 2007 found solid data to conclude that long-term exposure to PM2.5 (fine particulates) increases the overall risk of non-accidental mortality by 6% per a 10 microg/m3 increase. Exposure to PM2.5 was also associated with an increased risk of mortality from lung cancer (range: 15% to 21% per 10 microg/m3 increase) and total cardiovascular mortality (range: 12% to 14% per a 10 microg/m3 increase). The review further noted that living close to busy traffic appears to be associated with elevated risks of these three outcomes — increase in lung cancer deaths, cardiovascular deaths, and overall non-accidental deaths. The reviewers also found suggestive evidence that exposure to PM2.5 is positively associated with mortality from coronary heart diseases and exposure to SO_2 increases mortality from lung cancer, but the data was insufficient to provide solid conclusions. Another investigation showed that higher activity level increases deposition fraction of aerosol particles in human lung and recommended avoiding heavy activities like running in outdoor space at polluted areas.

In 2011, a large Danish epidemiological study found an increased risk of lung cancer for patients who lived in areas with high nitrogen oxide concentrations. In this study, the association was higher for non-smokers than smokers. An additional Danish study, also in 2011, likewise noted evidence of possible associations between air pollution and other forms of cancer, including cervical cancer and brain cancer.

Children

In the United States, despite the passage of the Clean Air Act in 1970, in 2002 at least 146 million Americans were living in non-attainment areas—regions in which the concentration of certain air pollutants exceeded federal standards. These dangerous pollutants are known as the criteria pollutants, and include ozone, particulate matter, sulfur dioxide, nitrogen dioxide, carbon monoxide, and lead. Protective measures to ensure children's health are being taken in cities such as New Delhi, India where buses now use compressed natural gas to help eliminate the "pea-soup" smog. A recent study in Europe has found that exposure to ultrafine particles can increase blood pressure in children.

HEALTH EFFECTS IN RELATIVELY "CLEAN" AREAS

Even in the areas with relatively low levels of air pollution, public health effects can be significant and costly, since a large number of people breathe in such pollutants. A 2005 scientific study for the British Columbia Lung Association showed that a small improvement in air quality (1% reduction of ambient PM2.5 and ozone concentrations) would produce $29 million in annual savings in the Metro Vancouver region in 2010. This finding is based on health valuation of lethal (death) and sub-lethal (illness) affects.

Effects on the Central Nervous System

Data is accumulating that air pollution exposure also affects the central nervous system.

In a June 2014 study conducted by researchers at the University of Rochester Medical Center, published in the journal Environmental Health Perspectives, it was discovered that early exposure to air pollution causes the same damaging changes in the brain as autism and schizophrenia. The study also shows that air pollution also affected short-term memory, learning ability, and impulsivity. Lead researcher Professor Deborah Cory-Slechta said that "When we looked closely at the ventricles, we could see that the white matter that normally surrounds them hadn't fully developed. It appears that inflammation had damaged those brain cells and prevented that region of the brain from developing, and the ventricles simply expanded to fill the space. Our findings add to the growing body of evidence that air pollution may play a role in autism, as well as in other neurodevelopmental disorders." Air pollution has a more significant negative effect of males than on females.

Agricultural Effects

In India in 2014, it was reported that air pollution by black carbon and ground level ozone had cut crop yields in the most affected areas by almost half in 2010 when compared to 1980 levels.

Historical Disasters

The world's worst short-term civilian pollution crisis was the 1984 Bhopal Disaster in India. Leaked industrial vapours from the Union Carbide factory, belonging to Union Carbide, Inc., U.S.A. (later bought by Dow Chemical Company), killed more than 25,000 people outright and injured anywhere from 150,000 to 600,000. The United Kingdom suffered its worst air pollution event when the December 4 Great Smog of 1952 formed over London. In six days more than 4,000 died and more recent estimates put the figure at nearer 12,000. An accidental leak of anthrax spores from a biological warfare laboratory in the former USSR in 1979 near Sverdlovsk is believed to have caused at least 64 deaths. The worst single incident of air pollution to occur in the US occurred in Donora, Pennsylvania in late October, 1948, when 20 people died and over 7,000 were injured.

Alternatives to Creating Air Pollution

There are now practical alternatives to the three principal causes of air pollution. Combustion of fossil fuels for space heating can be replaced by using ground source heat pumps and seasonal thermal energy storage. Electric power generation from burning fossil fuels can be replaced by power generation from nuclear and renewables. Motor vehicles driven by fossil fuels, a key factor in urban air pollution, can be replaced by electric vehicles.

Reduction Efforts

There are various air pollution control technologies and land-use planning strategies available to reduce air pollution. At its most basic level, land-use planning is likely to involve zoning and transport infrastructure planning. In most developed countries, land-use planning is an important part of social policy, ensuring that land is used efficiently for the benefit of the wider economy and population, as well as to protect the environment.

Because a large share of air pollution is caused by combustion of fossil fuels such as coal and oil, the reduction of these fuels can reduce air pollution drastically. Most effective is the switch to clean power sources such as wind power, solar power, hydro power which don't cause air pollution. Efforts to reduce pollution from mobile sources includes primary regulation (many developing countries have permissive regulations), expanding regulation to new sources (such as cruise and transport ships, farm equipment, and small gas-powered equipment such as string trimmers, chainsaws, and snowmobiles), increased fuel efficiency (such as through the use of hybrid vehicles), conversion to cleaner fuels or conversion to electric vehicles.

Titanium dioxide has been researched for its ability to reduce air pollution. Ultraviolet light will release free electrons from material, thereby creating free radicals, which break up VOCs and NOx gases. One form is superhydrophilic.

In 2014, Prof. Tony Ryan and Prof. Simon Armitage of University of Sheffield prepared a 10 meter by 20 meter-sized poster coated with microscopic, pollution-eating nanoparticles of titanium dioxide. Placed on a building, this giant poster can absorb the toxic emission from around 20 cars each day.

A very effective means to reduce air pollution is the transition to renewable energy. According to a study published in Energy and Environmental Science in 2015 the switch to 100% renewable energy in the United States would eliminate about 62,000 premature mortalities per year and about 42,000 in 2050, if no biomass were used. This would save about $600 billion in health costs a year due to reduced air pollution in 2050, or about 3.6% of the 2014 U.S. gross domestic product.

Control Devices

The following items are commonly used as pollution control devices in industry and transportation. They can either destroy contaminants or remove them from an exhaust stream before it is emitted into the atmosphere.

- Particulate control
- Mechanical collectors (dust cyclones, multicyclones)
- Electrostatic precipitators An electrostatic precipitator (ESP), or electrostatic air cleaner is a particulate collection device that removes particles from a flowing gas (such as air), using the force of an induced electrostatic charge. Electrostatic precipitators are highly efficient filtration devices that minimally impede the flow of gases through the device, and can easily remove fine particulates such as dust and smoke from the air stream.
- Baghouses Designed to handle heavy dust loads, a dust collector consists of a blower, dust filter, a filter-cleaning system, and a dust receptacle or dust

removal system (distinguished from air cleaners which utilize disposable filters to remove the dust).

- Particulate scrubbers Wet scrubber is a form of pollution control technology. The term describes a variety of devices that use pollutants from a furnace flue gas or from other gas streams. In a wet scrubber, the polluted gas stream is brought into contact with the scrubbing liquid, by spraying it with the liquid, by forcing it through a pool of liquid, or by some other contact method, so as to remove the pollutants.
- Scrubbers
- Baffle spray scrubber
- Cyclonic spray scrubber
- Ejector venturi scrubber
- Mechanically aided scrubber
- Spray tower
- Wet scrubber
- NOx control
- Low NOx burners
- Selective catalytic reduction (SCR)
- Selective non-catalytic reduction (SNCR)
- NOx scrubbers
- Exhaust gas recirculation
- Catalytic converter (also for VOC control)
- VOC abatement
- Adsorption systems, using activated carbon, such as Fluidized Bed Concentrator
- Flares
- Thermal oxidizers
- Catalytic converters
- Biofilters
- Absorption (scrubbing)
- Cryogenic condensers

- Vapor recovery systems
- Acid Gas/SO_2 control
- Wet scrubbers
- Dry scrubbers
- Flue-gas desulfurization
- Mercury control
- Sorbent Injection Technology
- Electro-Catalytic Oxidation (ECO)
- K-Fuel
- Dioxin and furan control
- Miscellaneous associated equipment
- Source capturing systems
- Continuous emissions monitoring systems (CEMS)

Regulations

In general, there are two types of air quality standards. The first class of standards (such as the U.S. National Ambient Air Quality Standards and E.U. Air Quality Directive) set maximum atmospheric concentrations for specific pollutants. Environmental agencies enact regulations which are intended to result in attainment of these target levels. The second class (such as the North American Air Quality Index) take the form of a scale with various thresholds, which is used to communicate to the public the relative risk of outdoor activity. The scale may or may not distinguish between different pollutants.

Canada

In Canada, air pollution and associated health risks are measured with the Air Quality Health Index or (AQHI). It is a health protection tool used to make decisions to reduce short-term exposure to air pollution by adjusting activity levels during increased levels of air pollution.

The Air Quality Health Index or "AQHI" is a federal program jointly coordinated by Health Canada and Environment Canada. However, the AQHI program would not be possible without the commitment and support of the provinces, municipalities and NGOs. From air quality monitoring to health risk communication and community engagement, local partners are responsible for the vast majority of work related to AQHI implementation. The AQHI provides a number from 1 to 10+ to indicate the level of health risk associated with local air quality. Occasionally, when the amount of air pollution is abnormally high, the number may exceed 10. The AQHI provides a local air quality current value as well as a local air quality maximums forecast for today, tonight and tomorrow and provides associated health advice.

Risk: Low (1-3) Moderate (4-6) High (7-10) Very high (above 10)

As it is now known that even low levels of air pollution can trigger discomfort for the sensitive population, the index has been developed as a continuum: The higher the number, the greater the health risk and need to take precautions. The index describes the level of health risk associated with this number as 'low', 'moderate', 'high' or 'very high', and suggests steps that can be taken to reduce exposure. The measurement is based on the observed relationship of Nitrogen Dioxide (NO2), ground-level Ozone (O_3) and particulates (PM2.5) with mortality, from an analysis of several Canadian cities. Significantly, all three of these pollutants can pose health risks, even at low levels of exposure, especially among those with pre-existing health problems.

When developing the AQHI, Health Canada's original analysis of health effects included five major air pollutants: particulates, ozone, and nitrogen dioxide (NO2), as well as sulfur dioxide (SO_2), and carbon monoxide (CO). The latter two pollutants provided little information in predicting health effects and were removed from the AQHI formulation.

The AQHI does not measure the effects of odour, pollen, dust, heat or humidity.

Germany

TA Luft is the German air quality regulation.

HOTSPOTS

Air pollution hotspots are areas where air pollution emissions expose individuals to increased negative health effects. They are particularly common in highly populated, urban areas, where there may be a combination of stationary sources (e.g. industrial facilities) and mobile sources (e.g. cars and trucks) of pollution. Emissions from these sources can cause respiratory disease, childhood asthma, cancer, and other health problems. Fine particulate matter such as diesel soot, which contributes to more than 3.2 million premature deaths around the world each year, is a significant problem. It is very small and can lodge itself within the lungs and enter the bloodstream. Diesel soot is concentrated in densely populated areas, and one in six people in the U.S. live near a diesel pollution hot spot.

While air pollution hotspots affect a variety of populations, some groups are more likely to be located in hotspots. Previous studies have shown disparities in exposure to pollution by race and/or income. Hazardous land uses (toxic storage and disposal facilities, manufacturing facilities, major roadways) tend to be located where property values and income levels are low. Low socioeconomic status can be a proxy for other kinds of social vulnerability, including race, a lack of ability to influence regulation and a lack of ability to move to neighborhoods with less environmental pollution. These communities bear a disproportionate burden of environmental pollution and are more likely to face health risks such as cancer or asthma.

Studies show that patterns in race and income disparities not only indicate a higher exposure to pollution but also higher risk of adverse health outcomes. Communities characterized by low socioeconomic status and racial minorities can be more vulnerable to cumulative adverse health impacts resulting from elevated exposure to pollutants than more privileged communities. Blacks and Latinos generally face more pollution than whites and Asians, and low-income communities bear a higher burden of risk than affluent ones. Racial discrepancies are particularly distinct in suburban areas of the US South and metropolitan areas of the US West. Residents in public housing, who are generally low-income and cannot move to healthier neighborhoods, are highly affected by nearby refineries and chemical plants.

Cities

Air pollution is usually concentrated in densely populated metropolitan areas, especially in developing countries where environmental regulations are relatively lax or nonexistent. However, even populated areas in developed countries attain unhealthy levels of pollution, with Los Angeles and Rome being two examples. Between 2002 and 2011 the incidence of lung cancer in Beijing near doubled. While smoking remains the leading cause of lung cancer in China, the number of smokers is falling while lung cancer rates are rising.

National-scale Air Toxics Assessments 1995-2005

The national-scale air toxics assessment(NATA) is an evaluation of air toxics by the U.S. EPA. EPA has furnished four assessments that characterize nationwide chronic cancer risk estimates and noncancer hazards from inhaling air toxics. The lates was from 2005, and made publicly available in early 2011.

"EPA developed the NATA as a state-of-the-science screening tool for State/Local/Tribal Agencies to prioritize pollutants, emission sources and locations of interest for further study, in order to gain a better understanding of the risks. NATA assessments do not incorporate refined information about emission sources, but rather, use general information about sources to develop estimates of risks which are more likely to overestimate impacts than underestimate them. NATA provides estimates of the risk of cancer and other serious health effects from breathing (inhaling) air toxics in order to inform both national and more localized efforts to identify and prioritize air toxics, emission source types and locations which are of greatest potential concern in terms of contributing to population risk. This in turn helps air pollution experts focus limited analytical resources on areas and or populations where the potential for health risks are highest. Assessments include estimates of cancer and non-cancer health effects based on chronic exposure from outdoor sources, including assessments of non-cancer health effects for Diesel Particulate Matter. Assessments provide a snapshot of the outdoor air quality and the risks to human health that would result if air toxic emissions levels remained unchanged."

Governing Urban Air Pollution

In Europe, Council Directive 96/62/EC on ambient air quality assessment and management provides a common strategy against which member states can "set objectives for ambient air quality in order to avoid, prevent or reduce harmful effects on human health and the environment . . . and improve air quality where it is unsatisfactory".

On 25 July 2008 in the case Dieter Janecek v Freistaat Bayern CURIA, the European Court of Justice ruled that under this directive citizens have the right to require national authorities to implement a short term action plan that aims to maintain or achieve compliance to air quality limit values.

This important case law appears to confirm the role of the EC as centralised regulator to European nation-states as regards air pollution control. It places a supranational legal obligation on the UK to protect its citizens from dangerous levels of air pollution, furthermore superseding national interests with those of the citizen.

In 2010, the European Commission (EC) threatened the UK with legal action against the successive breaching of PM10 limit values. The UK government has identified that if fines are imposed, they could cost the nation upwards of £300 million per year.

In March 2011, the Greater London Built-up Area remains the only UK region in breach of the EC's limit values, and has been given 3 months to implement an emergency action plan aimed at meeting the EU Air Quality Directive. The City of London has dangerous levels of PM10 concentrations, estimated to cause 3000 deaths per year within the city. As well as the threat of EU fines, in 2010 it was threatened with legal action for scrapping the western congestion charge zone, which is claimed to have led to an increase in air pollution levels.

In response to these charges, Boris Johnson, Mayor of London, has criticised the current need for European cities to communicate with Europe through their nation state's central government, arguing that in future "A great city like London" should be permitted to bypass its government and deal directly with the European Commission regarding its air quality action plan.

This can be interpreted as recognition that cities can transcend the traditional national government organisational hierarchy and develop solutions to air pollution using global governance networks, for example through transnational relations. Transnational relations include but are not exclusive to national governments and intergovernmental organisations, allowing sub-national actors including cities and regions to partake in air pollution control as independent actors.

Particularly promising at present are global city partnerships. These can be built into networks, for example the C40 Cities Climate Leadership Group, of which London is a member. The C40 is a public 'non-state' network of the world's leading cities that aims to curb their greenhouse emissions. The C40 has been identified as 'governance from the middle' and is an alternative to intergovernmental policy. It has the potential to improve urban air quality as participating cities "exchange information, learn from best practices and consequently mitigate carbon dioxide emissions independently from national government decisions". A criticism of the C40 network is that its exclusive nature limits influence to participating cities and risks drawing resources away from less powerful city and regional actors.

Atmospheric Dispersion

The basic technology for analyzing air pollution is through the use of a variety of mathematical models for predicting the transport of air pollutants in the lower atmosphere. The principal methodologies are:

- Point source dispersion, used for industrial sources

- Line source dispersion, used for airport and roadway air dispersion modeling
- Area source dispersion, used for forest fires or duststorms
- Photochemical models, used to analyze reactive pollutants that form smog

The point source problem is the best understood, since it involves simpler mathematics and has been studied for a long period of time, dating back to about the year 1900. It uses a Gaussian dispersion model for continuous buoyant pollution plumes to predict the air pollution isopleths, with consideration given to wind velocity, stack height, emission rate and stability class (a measure of atmospheric turbulence). This model has been extensively validated and calibrated with experimental data for all sorts of atmospheric conditions.

The roadway air dispersion model was developed starting in the late 1950s and early 1960s in response to requirements of the National Environmental Policy Act and the U.S. Department of Transportation (then known as the Federal Highway Administration) to understand impacts of proposed new highways upon air quality, especially in urban areas. Several research groups were active in this model development, among which were: the Environmental Research and Technology (ERT) group in Lexington, Massachusetts, the ESL Inc. group in Sunnyvale, California and the California Air Resources Board group in Sacramento, California. The research of the ESL group received a boost with a contract award from the United States Environmental Protection Agency to validate a line source model using sulfur hexafluoride as a tracer gas. This program was successful in validating the line source model developed by ESL Inc. Some of the earliest uses of the model were in court cases involving highway air pollution; the Arlington, Virginia portion of Interstate 66 and the New Jersey Turnpike widening project through East Brunswick, New Jersey.

Area source models were developed in 1971 through 1974 by the ERT and ESL groups, but addressed a smaller fraction of total air pollution emissions, so that their use and need was not as widespread as the line source model, which enjoyed hundreds of different applications as early as the 1970s. Similarly photochemical models were developed primarily in the 1960s and 70s, but their use was more specialized and for regional needs, such as understanding smog formation in Los Angeles, California.

Air Pollution and Environmental Justice

No one wants to live next door to an incinerator, oil refinery, port, toxic waste dump, or other polluting site. Yet millions of people around the world do, and this puts them at a much higher risk for respiratory disease, cardiovascular disease, neurological damage, cancer, and death. In the United States, people of color are 1.5 times more likely than whites to live in areas with poor air quality, according to the ALA.

Historically, racist zoning policies and the discriminatory lending practices known as redlining have combined to keep polluting industries and car-choked highways away from white neighborhoods and have turned communities of color—especially poor and working-class communities of color—into sacrifice zones where residents are forced to breathe dirty air and suffer the many health problems associated with it. In addition to the increased health risks that come from living in such places, members of these communities experience economic harm in the form of missed workdays, higher medical costs, and local underinvestment.

Environmental racism isn't limited to cities and industrial areas. Outdoor laborers, including the estimated three million migrant and seasonal farmworkers in the United States, are among the most vulnerable to air pollution—and also among the least equipped, politically, to pressure employers and lawmakers to affirm their right to breathe clean air.

Recently, cumulative impact mapping, which uses data on environmental conditions and demographics, has been able to show how some communities are overburdened with layers of issues, like high levels of poverty, unemployment, and pollution. Tools like the Environmental Justice Screening Method and the EPA's EJSCREEN provide evidence of what many environmental justice communities have been explaining for decades: that we need land-use and public health reforms to ensure that vulnerable areas are not overburdened and that the people who need resources most are receiving them.

Controlling Air Pollution

In the United States, the Clean Air Act has been a crucial tool for reducing air pollution since its passage in 1970, although fossil-fuel interests aided by industry-friendly lawmakers have frequently attempted to weaken its many protections. Ensuring that this bedrock environmental law remains intact and properly enforced will always be key to maintaining and improving our air quality.

But the best, most effective way to control air pollution is to speed up our transition to cleaner fuels and industrial processes. By switching over to renewable energy sources (such as wind and solar power), maximizing fuel efficiency in our vehicles, and replacing more and more of our gasoline-powered cars and trucks with electric versions, we'll be limiting air pollution at its source while also curbing the global warming that heightens so many of its worst health impacts.

And what about the economic costs of controlling air pollution? According to a report on the Clean Air Act commissioned by NRDC, the annual benefits of cleaner air are up to 32 times greater than the cost of clean-air regulations. Those benefits include up to 370,000 avoided premature deaths, 189,000 fewer hospital admissions for cardiac and respiratory illnesses, and net economic benefits of up to $3.8 trillion for the U.S. economy every year.

GLOBAL BURDEN OF AIR POLLUTION

Ambient (outdoor) and household (indoor) air pollution cause an estimated 7 million premature deaths each year,22 including approximately 600,000 deaths per year in children under 5 years of age.23 Outside the home, incineration and combustion of fossil fuels emit a complex array of pollutants, including particulate matter, nitrogen dioxide, sulphur dioxide and carbon monoxide, which may react in the air to generate secondary pollutants, such as ground-level ozone. In 2016, 91% of the global population lived in places that did not meet the World Health Organization's (WHO) air quality guideline levels for particulate matter,4 and more than 50% lived in cities with at least 2.5 times the recommended levels of fine particulate matter.24 The latest air quality data from over 100 countries can be viewed on the WHO interactive map.25

Impact of Air Pollution On Health

The effects of ambient (outdoor) and household (indoor) air pollution on human health are wide-ranging. Inhalation of airborne pollutants damages the respiratory tract, reduces lung function, and causes or exacerbates acute and chronic respiratory diseases, including asthma.5 More than 25% of premature deaths associated with air pollution are respiratory in nature.12

Outdoor air pollution and particulate matter are classed as carcinogenic to humans29 and exposure to air pollution is estimated to cause 29% of all adult deaths from lung cancer.2 It is estimated that air pollution causes 43% of all adult deaths from chronic obstructive pulmonary disease, 25% from stroke and 24% from heart disease.2 Household air pollution is responsible for nearly half (45%) of all pneumonia deaths in children under 5 years of age and contributes to 28% of pneumonia deaths in adults.3 There is mounting evidence for the adverse effects of prenatal exposure to air pollution on lung development, respiratory health and development of chronic disease in adulthood.30 BreatheLife – a joint campaign led by the World Health Organization (WHO), United Nations Environment, and Climate & Clean Air Coalition – provides tips on reducing prenatal exposure.31

In recent years, evidence for effects of inhaled pollutants on the central nervous system has also emerged. Exposure to air pollution has been associated with impaired cognitive and psychomotor development in children, and with cognitive decline and dementia in the elderly.32

Click here to learn more about the mechanisms through which air pollution leads to adverse health effects.

Why breathing clean air is important

Life-time exposure to ambient air pollution, in a highly polluted city such as Beijing in China, may result in an estimated loss of life expectancy of approximately 3 years.33,34 Other estimates suggest that long-term exposure to air pollution may reduce life expectancy by as much as 11 years (based on WHO consensus estimates for the mortality risk ratio related to an annual PM2.5 increase of 10 μg/m3).35

By reducing air pollution in cities to WHO air quality guideline levels (PM2.5 annual mean concentrations reduced to 10 μg/m3), air pollution-related deaths could be reduced by 15%.4

The Energy Policy Institute at the University of Chicago have estimated the extra life gained per person if air pollution in major cities met WHO air quality guideline levels:36

Ambient air pollution

Exposure to ambient air pollution was estimated to cause 4.2 million premature deaths worldwide in 2016.4 In 2015, the Global Burden of Diseases study identified ambient PM2.5 (particulate matter with diameter <2.5 μm) as responsible for 7.6% of total global mortality - this makes it the fifth most important global risk factor for death.26

Ambient air pollution affects all geographical regions and countries, and people of all socioeconomic groups.22 However, low- and middle-income (LMI) countries are the most affected, with the poorest air quality in Eastern Mediterranean and African regions.24 In 2016, 91% of deaths linked to ambient air pollution occurred in LMI countries, with the highest numbers in South-East Asia and Western Pacific regions.4

Household air pollution

Household air pollution is another major cause of premature death. Around 3 billion people still cook using solid fuels or kerosene in inefficient stoves or open hearths.3 This releases particulate matter, methane, carbon monoxide, polyaromatic hydrocarbons and volatile organic compounds into the home,27 and also contributes to ambient air pollution. Each year, 3.8 million people die prematurely as a result of household air pollution.3

Household air pollution from cooking fuels is strongly linked to poverty, with rural households in Sub-Saharan Africa, South and East Asia and the Western Pacific experiencing the greatest health burden.27 This interactive graph28 shows mortality from household and ambient air pollution in WHO regions and countries.

How to Help Reduce Air Pollution

"The less gasoline we burn, the better we're doing to reduce air pollution and harmful effects of climate change," Walke says. "Make good choices about transportation. When you can, walk, ride a bike, or take public transportation. For driving, choose a car that gets better miles per gallon of gas, or choose an electric car." You can also investigate your power provider options—you may be able to request that your electricity be supplied by wind or solar. Buying your food locally cuts down on the fossil fuels burned in trucking or flying food in from across the country. And most important, "Support leaders who push for clean air and water and responsible steps on climate change," Walke says.

How to Protect Your Health

"When you see in the news or hear on the weather report that pollution levels are high, it may be useful to limit the time when children go outside or you go for a jog," Walke says. Generally, ozone levels tend to be lower in the morning.

If you exercise outside, stay as far as you can from heavily trafficked roads. Then shower and wash your clothes to remove fine particles.

The air may look clear, but that doesn't mean it's pollution free. Utilize tools like the EPA's air pollution monitor, AirNow, to get the latest conditions. If the air quality is bad, stay inside with windows closed.

If you live or work in an area prone to wildfires, stay away from the harmful smoke as much as you're able. Consider keeping a small stock of masks to wear when conditions are poor.

Wear sunscreen. When ultraviolet radiation comes through the weakened ozone layer, it can cause skin damage and skin cancer.

Acid Rain

Acid rain, or acid deposition, is a broad term that includes any form of precipitation with acidic components, such as sulfuric or nitric acid that fall to the ground from the atmosphere in wet or dry forms. This can include rain, snow, fog, hail or even dust that is acidic.

Acid rain results when sulfur dioxide (SO_2) and nitrogen oxides (NOX) are emitted into the atmosphere and transported by wind and air currents. The SO_2 and NOX react with water, oxygen and other chemicals to form sulfuric and nitric acids. These then mix with water and other materials before falling to the ground.

While a small portion of the SO_2 and NOX that cause acid rain is from natural sources such as volcanoes, most of it comes from the burning of fossil fuels. The major sources of SO_2 and NOX in the atmosphere are:

Burning of fossil fuels to generate electricity. Two thirds of SO2 and one fourth of NOX in the atmosphere come from electric power generators.

Vehicles and heavy equipment.

Manufacturing, oil refineries and other industries.

Winds can blow SO_2 and NOX over long distances and across borders making acid rain a problem for everyone and not just those who live close to these sources.

Acid rain, also called acid precipitation or acid deposition, precipitation possessing a pH of about 5.2 or below primarily produced from the emission of sulfur dioxide (SO_2) and nitrogen oxides (NOx; the combination of NO and NO_2) from human activities, mostly the combustion of fossil fuels. In acid-sensitive landscapes, acid deposition can reduce the pH of surface waters and lower biodiversity. It weakens trees and increases their susceptibility to damage from other stressors, such as drought, extreme cold, and pests. In acid-sensitive areas, acid rain also depletes soil of important plant nutrients and buffers, such as calcium and magnesium, and can release aluminum, bound to soil particles and rock, in its toxic dissolved form. Acid rain contributes to the corrosion of surfaces exposed to air pollution and is responsible for the deterioration of limestone and marble buildings and monuments.

The phrase acid rain was first used in 1852 by Scottish chemist Robert Angus Smith during his investigation of rainwater chemistry near industrial cities in England and Scotland. The phenomenon became an important part of his book Air and Rain: The Beginnings of a Chemical Climatology (1872). It was not until the late 1960s and early 1970s, however, that acid rain was recognized as a regional environmental issue affecting large areas of western Europe and eastern North America. Acid rain also occurs in Asia and parts of Africa, South America, and Australia. As a global environmental issue, it is frequently overshadowed by climate change. Although the problem of acid rain has been significantly reduced in some areas, it remains an important environmental issue within and downwind from major industrial and industrial agricultural regions worldwide.

FORMS OF ACID DEPOSITION

Wet Deposition

Wet deposition is what we most commonly think of as acid rain. The sulfuric and nitric acids formed in the atmosphere fall to the ground mixed with rain, snow, fog, or hail.

Dry Deposition

Acidic particles and gases can also deposit from the atmosphere in the absence of moisture as dry deposition. The acidic particles and gases may deposit to surfaces (water bodies, vegetation, buildings) quickly or may react during atmospheric transport to form larger particles that can be harmful to human health. When the accumulated acids are washed off a surface by the next rain, this acidic water flows over and through the ground, and can harm plants and wildlife, such as insects and fish.

The amount of acidity in the atmosphere that deposits to earth through dry deposition depends on the amount of rainfall an area receives. For example, in desert areas the ratio of dry to wet deposition is higher than an area that receives several inches of rain each year.

Measuring Acid Rain

A diagram showing where various substances fall on the pH scale.Acidity and alkalinity are measured using a pH scale for which 7.0 is neutral. The lower a substance's pH (less than 7), the more acidic it is; the higher a substance's pH (greater than 7), the more alkaline it is. Normal rain has a pH of about 5.6; it is slightly acidic because carbon dioxide (CO_2) dissolves into it forming weak carbonic acid. Acid rain usually has a pH between 4.2 and 4.4.

Policymakers, research scientists, ecologists, and modelers rely on the National Atmospheric Deposition Program's (NADP) National Trends Network (NTN) for measurements of wet deposition. The NADP/NTN collects acid rain at more than 250 monitoring sites throughout the US, Canada, Alaska, Hawaii and the US Virgin Islands. Unlike wet deposition, dry deposition is difficult and expensive to measure. Dry deposition estimates for nitrogen and sulfur pollutants are provided by the Clean Air Status and Trends Network (CASTNET). Air concentrations are measured by CASTNET at more than 90 locations.

When acid deposition is washed into lakes and streams, it can cause some to turn acidic. The Long-Term Monitoring (LTM) Network measures and monitors surface water chemistry at over 280 sites to provide valuable information on aquatic ecosystem health and how water bodies respond to changes in acid-causing emissions and acid deposition.

Chemistry of acid deposition

Acid rain is a popular expression for the more scientific term acid deposition, which refers to the many ways in which acidity can move from the atmosphere to Earth's surface. Acid deposition includes acidic rain as well as other forms of acidic wet deposition—such as snow, sleet, hail, and fog (or cloud water). Acid deposition also includes the dry deposition of acidic particles and gases, which can affect landscapes during dry periods. Thus, acid deposition is capable of affecting landscapes and the living things that reside within them even when precipitation is not occurring.

Acidity is a measure of the concentration of hydrogen ions (H^+) in a solution. The pH scale measures whether a solution is acidic or basic. Substances are considered acidic below a pH of 7, and each unit of pH below 7 is 10 times more acidic, or has 10 times more H+, than the unit above it. For example, rainwater with a pH of 5.0 has a concentration of 10 microequivalents of H^+ per litre, whereas rainwater with a pH of 4.0 has a concentration of 100 microequivalents of H^+ per litre.

Normal rainwater is weakly acidic because of the absorption of carbon dioxide (CO_2) from the atmosphere—a process that produces carbonic acid—and from organic acids generated from biological activity. In addition, volcanic activity can produce sulfuric acid (H_2SO_4), nitric acid (HNO_3), and hydrochloric acid (HCl) depending on the emissions associated with specific volcanoes. Other natural sources of acidification include the production of nitrogen oxides from the conversion of atmospheric molecular nitrogen (N_2) by lightning and the conversion of organic nitrogen by wildfires. However, the geographic extent of any given natural source of acidification is small, and in most cases it lowers the pH of precipitation to no more than about 5.2. Anthropogenic activities, particularly the burning of fossil fuels (coal, oil, natural gas) and the smelting of metal ores, are the major causes of acid deposition. In the United States, electric utilities produce nearly 70 percent of SO_2 and about 20 percent of NOx emissions. Fossil fuels burned by vehicles account for nearly 60 percent of NOx emissions in the United States. In the atmosphere, sulfuric and nitric acids are generated when SO_2 and NOx, respectively, react with water. The simplest reactions are:

SO_2 and NOx emissions in the U.S.

SO_2 and NOx emissions in the U.S.

SO_2 and NOx emissions in the U.S., 2008.

Encyclopædia Britannica, Inc.

$SO_2 + H_2O$? H_2SO_4 ?? H+ + HSO_4 ?? 2H+ + SO_{42}

$NO_2 + H_2O$? HNO_3 ?? H+ + NO_3

These reactions in the aqueous phase (for example, in cloud water) create wet deposition products. In the gaseous phase they can produce acidic dry deposition. Acid formation can also occur on particles in the atmosphere.

Where fossil fuel consumption is large and emission controls are not in place to reduce SO_2 and NOx emissions, acid deposition will occur in areas downwind of emission sources, often hundreds to thousands of kilometres away. In such areas the pH of precipitation can average 4.0 to 4.5 annually, and the pH of individual rain events can sometimes drop below 3.0. In addition, cloud water and fog in polluted areas may be many times more acidic than rain falling over the same region.

Many air pollution and atmospheric deposition problems are intertwined with one another, and these problems are often derived from the same cause, namely the burning of fossil fuels. In addition to acid deposition, NOx emissions along with hydrocarbon emissions are key ingredients in ground-level ozone (photochemical smog) formation, which is one of the most widespread forms of air pollution. The SO_2 and NOx emissions can generate fine particulates, which are harmful to human respiratory systems. Coal combustion is the leading source of atmospheric mercury, which also enters ecosystems by wet and dry deposition. (A number of other heavy metals, such as lead and cadmium, and various particulates are also products of unregulated fossil fuel combustion.) Acid deposition of nitrogen derived from NOx emissions creates additional environmental problems. For example, many lake, estuarine, and coastal marine systems receive too much nitrogen from atmospheric deposition and terrestrial runoff. This eutrophication (or over-enrichment) causes the overgrowth of plants and algae. When these organisms die and decompose, they deplete the dissolved oxygen supply necessary for most aquatic life in water bodies. Eutrophication is considered to be a major environmental problem in lake, coastal marine, and estuarine ecosystems worldwide.

ECOLOGICAL EFFECTS OF ACID DEPOSITION

Effects on lakes and rivers

The regional effects of acid deposition were first noted in parts of western Europe and eastern North America in the late 1960s and early 1970s when changes in the chemistry of rivers and lakes, often in remote locations, were linked to declines in the health of aquatic organisms such as resident fish, crayfish, and clam populations. Increasing amounts of acid deposition in sensitive areas caused tens of thousands of lakes and streams in Europe and North America to become much more acidic than they had been in previous decades. Acid-sensitive areas are those that are predisposed to acidification because the region's soils have a low buffering capacity, or low acid-neutralizing capacity (ANC). In addition, acidification can release aluminum bound to soils, which in its dissolved form can be toxic to both plant and animal life. High concentrations of dissolved aluminum released from soils often enter streams and lakes. In conjunction with rising acidity in aquatic environments, aluminum can damage fish gills and thus impair respiration. In the Adirondack Mountain region of New York state, research has shown that the number of fish species drops from five in lakes with a pH of 6.0 to 7.0 to only one in lakes with a pH of 4.0 to 4.5. Other organisms are also negatively affected, so that acidified bodies of water lose plant and animal diversity overall. These effects can ripple throughout the food chain.

High acidity, especially from sulfur deposition, can accelerate the conversion of elemental mercury to its deadliest form: methyl mercury, a neurological toxin. This conversion most commonly occurs in wetlands and water-saturated soils where low-oxygen environments provide ideal conditions for the formation of methyl mercury by bacteria. Methyl mercury concentrates in organisms as it moves up the food chain, a phenomenon known as bioaccumulation. Small concentrations of methyl mercury present in phytoplankton and zooplankton accumulate in the fat cells of the animals that consume them. Since animals at higher tiers of the food chain must always consume large numbers of organisms from lower ones, the concentrations of methyl mercury in top predators, which often include humans, increase to levels where they could become harmful. The bioaccumulation of methyl mercury in the tissues of fishes is the leading reason for government health advisories that recommend reduced consumption of fish from fresh and marine waters.

In addition, aquatic acidification may be episodic, especially in colder climates. Sulfuric and nitric acid accumulating in a snowpack can leach out rapidly during the initial snowmelt and result in a pulse of acidic meltwater. Such pulses may be much more acidic than any individual snowfall event over the course of a winter, and these events can be deadly to acid-sensitive aquatic organisms throughout the food web.

Effects on forested and mountainous regions

In the 1970s and '80s, forested areas in central Europe, southern Scandinavia, and eastern North America showed alarming signs of forest dieback and tree mortality. A 1993 survey in 27 European countries revealed air pollution damage or mortality in 23 percent of the 100,000 trees surveyed. It is likely that the dieback was the result of many factors, including acid deposition (e.g., soil acidification and loss of buffering capacity, mobilization of toxic aluminum, direct effects of acid on foliage), exposure to ground-level ozone, possible excess fertilization from the deposition of nitrogen compounds (such as nitrates, ammonium, and ammonia compounds), and general stress caused by a combination of these factors. Once a tree is in a weakened condition, it is more likely to succumb to other environmental stressors such as drought, insect infestation, and infection by pathogens. The areas of forest dieback were often found to be associated with regions with low buffering capacity where damage to aquatic ecosystems due to acid deposition was also occurring.

The emission of sulfur dioxide and nitrogen oxides to the atmosphere by human activities—primarily fossil-fuel burning—has led to the acidification...

Acid deposition has been implicated in the alteration of soil chemistry and the decline of several tree species through both direct and indirect means. Poorly buffered soils are particularly susceptible to acidification because they lack significant amounts of base cations (positively charged ions), which neutralize acidity. Calcium, magnesium, sodium, and potassium, which are the base cations that account for most of the acid-neutralizing capacity of soils, are derived from the weathering of rocks and from wet and dry deposition. Some of these base cations (such as calcium and magnesium) are also secondary plant nutrients that are necessary for proper plant growth. The supply of these base cations declines as they neutralize the acids present in wet and dry deposition and are leached from the soils. Thus, a landscape formerly rich in base cations can become acid-sensitive when soil-formation processes are slow and base cations are not replaced through weathering or deposition processes.

Soil acidification can also occur where deposition of ammonia (NH_3) and ammonium (NH_4^+) is high. Ammonia and ammonium deposition leads to the production of H+ (which results in acidification) when these chemicals are converted to nitrate (NO3-) by bacteria in a process called nitrification:

$NH_3 + O_2$? $NO_2^- + 3H^+ + 2e.$

$NO_2^- + H_2O$? $NO_3^- + 2H^+ + 2e.$

The sources of NH_3 and NH_4^+ are largely agricultural activities, especially livestock (chickens, hogs, and cattle) production. Around 80 percent of NH_3 emissions in the United States and Europe come from the agricultural sector. The evaporation or volatilization of animal wastes releases NH_3 into the atmosphere. This process often results in the deposition of ammonia near the emission source.

However, NH_3 can be converted to particulate ammonium that may be transported and deposited as wet and dry deposition hundreds of kilometres away from the emission source.

Besides negatively altering soil chemistry, acid deposition has been shown to affect some tree species directly. Red spruce (*Picea rubens*) trees found at higher elevations in the eastern United States are harmed by acids leaching calcium from the cell membranes in their needles, making the needles more susceptible to damage from freezing during winter. The damage is often greatest in mountainous regions, because these areas often receive more acid deposition than lower areas and the winter environment is more extreme. Mountainous regions are subjected to highly acidic cloud and fog water along with other environmental stresses. In addition, red spruce can be damaged by the increased concentration of toxic aluminum in the soil. These processes can reduce nutrient uptake by the tree roots. Sugar maple (Acer saccharum) populations are also declining in the northeastern United States and parts of eastern Canada. High soil aluminum and low soil calcium concentrations resulting from acid deposition have been implicated in this decline. Other trees in this region that are negatively affected by acidic deposition include aspen (Populus), birch (Betula), and ash (Fraxinus).

Some scientists argue that acid deposition may influence the geology of some regions. A 2018 study examining the 2009 Jiweishan landslide in southwest China proposed that acid rain may have weakened a layer of shale that separated the rock layers containing an aquifer above from the rock layers containing a mine below, which caused a large mass of rock to slip off the mountainside and kill 74 people.

EFFECTS ON HUMAN-MADE STRUCTURES

Acid deposition also affects human-made structures. The most notable effects occur on marble and limestone, which are common building materials found in many historic structures, monuments, and gravestones. Sulfur dioxide, an acid rain precursor, can react directly with limestone in the presence of water to form gypsum, which eventually flakes off or is dissolved by water. In addition, acid rain can dissolve limestone and marble through direct contact.

The Effects of Acid Rain on Ecosystems

Graph showing level of acidity that is tolerable to various species of aquatic life

This figure illustrates the pH level at which key organisms may be lost as their environment becomes more acidic. Not all fish, shellfish, or the insects that they eat can tolerate the same amount of acid. An ecosystem is a community of plants, animals and other organisms along with their environment including the air, water and soil. Everything in an ecosystem is connected. If something harms one part of an ecosystem – one species of plant or animal, the soil or the water – it can have an impact on everything else.

Effects of Acid Rain on Fish and Wildlife

The ecological effects of acid rain are most clearly seen in aquatic environments, such as streams, lakes, and marshes where it can be harmful to fish and other wildlife. As it flows through the soil, acidic rain water can leach aluminum from soil clay particles and then flow into streams and lakes. The more acid that is introduced to the ecosystem, the more aluminum is released.

Some types of plants and animals are able to tolerate acidic waters and moderate amounts of aluminum. Others, however, are acid-sensitive and will be lost as the pH declines. Generally, the young of most species are more sensitive to environmental conditions than adults. At pH 5, most fish eggs cannot hatch. At lower pH levels, some adult fish die. Some acidic lakes have no fish. Even if a species of fish or animal can tolerate moderately acidic water, the animals or plants it eats might not. For example, frogs have a critical pH around 4, but the mayflies they eat are more sensitive and may not survive pH below 5.5.

Effects of Acid Rain on Plants and Trees

Dead or dying trees are a common sight in areas effected by acid rain. Acid rain leaches aluminum from the soil. That aluminum may be harmful to plants as well as animals. Acid rain also removes minerals and nutrients from the soil that trees need to grow. At high elevations, acidic fog and clouds might strip nutrients from trees' foliage, leaving them with brown or dead leaves and needles. The trees are then less able to absorb sunlight, which makes them weak and less able to withstand freezing temperatures.

Buffering Capacity

Many forests, streams, and lakes that experience acid rain don't suffer effects because the soil in those areas can buffer the acid rain by neutralizing the acidity in the rainwater flowing through it. This capacity depends on the thickness and composition of the soil and the type of bedrock underneath it. In areas such as mountainous parts of the Northeast United States, the soil is thin and lacks the ability to adequately neutralize the acid in the rain water. As a result, these areas are particularly vulnerable and the acid and aluminum can accumulate in the soil, streams, or lakes.

EPISODIC ACIDIFICATION

Melting snow and heavy rain downpours can result in what is known as episodic acidification. Lakes that do not normally have a high level of acidity may temporarily experience effects of acid rain when the melting snow or downpour brings greater amounts of acidic deposition and the soil can't buffer it. This short duration of higher acidity (i.e., lower pH) can result in a short-term stress on the ecosystem where a variety of organisms or species may be injured or killed.

Nitrogen Pollution

It's not just the acidity of acid rain that can cause problems. Acid rain also contains nitrogen, and this can have an impact on some ecosystems. For example, nitrogen pollution in our coastal waters is partially responsible for declining fish and shellfish populations in some areas. In addition to agriculture and wastewater, much of the nitrogen produced by human activity that reaches coastal waters comes from the atmosphere.

Effects of Acid Rain on Materials

Not all acidic deposition is wet. Sometimes dust particles can become acidic as well, and this is called dry deposition. When acid rain and dry acidic particles fall to earth, the nitric and sulfuric acid that make the particles acidic can land on statues, buildings, and other manmade structures, and damage their surfaces. The acidic particles corrode metal and cause paint and stone to deteriorate more quickly. They also dirty the surfaces of buildings and other structures such as monuments.

The consequences of this damage can be costly: damaged materials that need to be repaired or replaced, increased maintenance costs, and loss of detail on stone and metal statues, monuments and tombstones.

OTHER EFFECTS OF SO_2 AND NOX

Visibility

In the atmosphere, SO_2 and NOX gases can be transformed into sulfate and nitrate particles, while some NOX can also react with other pollutants to form ozone. These particles and ozone make the air hazy and difficult to see through. This affects our enjoyment of national parks that we visit for the scenic view such as Shenandoah and the Great Smoky Mountains.

Human Health

Walking in acid rain, or even swimming in a lake affected by acid rain, is no more dangerous to humans than walking in normal rain or swimming in non-acidic lakes. However, when the pollutants that cause acid rain —SO_2 and NOX, as well as sulfate and nitrate particles— are in the air, they can be harmful to humans.

SO_2 and NOX react in the atmosphere to form fine sulfate and nitrate particles that people can inhale into their lungs. Many scientific studies have shown a relationship between these particles and effects on heart function, such as heart attacks resulting in death for people with increased heart disease risk, and effects on lung function, such as breathing difficulties for people with asthma.

Solutions to Acid Rain

1. Cleaning up Exhaust Pipes and Smokestacks

Most of the electric power supporting the modern-day energy requirements comes from combusting fossil fuels such as oil, natural gas, and coal that generate nitrogen oxides (NOx) and sulfur dioxide (SO_2) as the chief contributors to acid rain. Burning coal largely accounts for SO_2 emissions while NOx emissions are mostly from fossil fuel combustions.

Washing coal, use of coal comprised of low sulfur, and use of devices known as "scrubbers" can provide a technical solutions to SO_2 emissions. "Scrubbing" also called flue-gas desulfurization (FGD) typically works to chemically eliminate SO_2 from the gases leaving smokestacks.

It can eliminate up to 95% of SO_2 gases. Power generation facilities can also shift to using fuels that emit much less SO_2 such as natural gas instead of burning coal. These methods are simply called emission reduction strategies.

Similarly, NOx emissions from automobile fossil fuel combustions are mitigated upon by the use of catalytic converters. Catalytic converters are fixed on the exhaust pipe system to reduce NOx emission. Improvement of gasoline that combusts cleaner is also a strategy for reducing the emission of NOx gases.

2. Restoring Damaged Environments

Use of limestone or lime, a process called liming, is a practice that people can do to repair the damage caused by acid rain to lakes, rivers and brooks. Adding lime into acidic surface waters balances the acidity. It's a process that has extensively been used, for instance in Sweden, to keep the water pH at optimum.

Even though, liming is an expensive method and has to be done repeatedly. Furthermore, it only offers a short-term solution at the expense of solving the broader challenges of SO_2 and NOx emissions and risks to human health. Nevertheless, it helps to restore and allow the survival of aquatic life forms by improving chronically acidified surface waters.

3. Alternative Energy Sources

Besides fossil fuels, there is a wide range of alternative energy sources that can generate electrical power. These include wind energy, geothermal energy, solar energy, hydropower, and nuclear power.

Harnessing these energy sources can offer effective electrical power alternatives instead of using fossil fuels. Fuel cells, natural gas, and batteries can also substitute the use of fossil fuel as cleaner energy sources. As of today, all energy sources have environmental and economic costs as well as benefits. The only solution is using sustainable energy that can protect the future.

INDIVIDUAL, NATIONAL/STATE, AND INTERNATIONAL ACTIONS

Millions of people, directly and indirectly, contribute to SO_2 and NOx emissions. Mitigation of this challenge requires individuals to be more informed about energy conservation and ways of reducing emissions such as: turning off lights or electrical appliances when not using them; use public transport; use energy-efficient electrical appliances; and use of hybrid vehicles or those with low NOx emissions.

Smog Effect

The smog effect or beret effect happens when there is a kind of dark fog concentrated over the cities and fields. That fog is a load of pollutants and can be of 2 types: sulphurous smog and photochemical smog, both dangerous and harmful to health. Both types of smog are a consequence of industrial and urban action. However, sulphurous smog has its origin mainly in the use of coal in many industrial processes. That has been reduced in the most advanced countries – nevertheless, there are still many developing countries that do not have treatment protocols for pollutants. Smog is a type of air pollutant. The word "smog" was made in the early 20th century as a portmanteau of the words smoke and fog to refer to smoky fog. The word was then intended to refer to what was sometimes known as pea soup fog, a familiar and serious problem in London from the 19th century to the mid 20th century. This kind of smog is caused by the burning of large amounts of coalwithin a city; this smog contains soot particulates from smoke,sulfur dioxide and other components. Modern smog, as found for example in Los Angeles, is a type of air pollution derived from vehicular emission from internal combustion engines and industrial fumes that react in the atmosphere with sunlight to form secondary pollutants that also combine with the primary emissions to form photochemical smog. In certain other cities, such as Delhi, smog severity is often aggravated by stubble burning in neighboring agricultural areas. The atmospheric pollution levels of Los Angeles, Beijing, Delhi, Mexico City and other cities are increased by inversion that traps pollution close to the ground. It is usually highly toxic to humans and can cause severe sickness, shortened life or death.

Etymology

Coinage of the term "smog" is generally attributed to Dr. Henry Antoine Des Voeux in his 1905 paper, "Fog and Smoke" for a meeting of the Public Health Congress. The July 26, 1905 edition of the London newspaper Daily Graphic quoted Des Voeux, "He said it required no science to see that there was something produced in great cities which was not found in the country, and that was smoky fog, or what was known as 'smog.'" The following day the newspaper stated that "Dr. Des Voeux did a public service in coining a new word for the London fog." "Smog" also appears in a January 19, 1893, Los Angeles Timesarticle and is attributed to "a witty English writer."

Causes

Coal

Coal fires, used to heat individual buildings or in a power-producing plant, can emit significant clouds of smoke that contributes to smog. Air pollution from this source has been reported in England since the Middle Ages. London, in particular, was notorious up through the mid-20th century for its coal-caused smogs, which were nicknamed 'pea-soupers.' Air pollution of this type is still a problem in areas that generate significant smoke from burning coal, as witnessed by the 2013 autumnal smog in Harbin, China, which closed roads, schools, and the airport.

Transportation emissions

Traffic emissions – such as from trucks, buses, and automobiles – also contribute. Airborne by-productsfrom vehicle exhaust systems cause air pollution and are a major ingredient in the creation of smog in some large cities. The major culprits from transportation sources are carbon monoxide (CO), nitrogen oxides (NO and NOx), volatile organic compounds, sulfur dioxide, and hydrocarbons. These molecules react with sunlight, heat, ammonia, moisture, and other compounds to form the noxious vapors, ground level ozone, and particles that comprise smog.

Photochemical smog

Photochemical smog was first described in the 1950s. It is the chemical reaction of sunlight, nitrogen oxidesand volatile organic compounds in the atmosphere, which leaves airborne particles and ground-level ozone. This noxious mixture of air pollutants can include the following:

- Aldehydes
- Nitrogen oxides, such as nitrogen dioxide
- Peroxyacyl nitrates
- Tropospheric ozone
- Volatile organic compounds

All of these harsh chemicals are usually highly reactive and oxidizing. Photochemical smog is therefore considered to be a problem of modern industrialization. It is present in all modern cities, but it is more common in cities with sunny, warm, dry climates and a large number of motor vehicles. Because it travels with the wind, it can affect sparsely populated areas as well.

NATURAL CAUSES

An erupting volcano can also emit high levels of sulphur dioxide along with a large quantity of particulate matter; two key components to the creation of smog. However, the smog created as a result of a volcanic eruption is often known as vog to distinguish it as a natural occurrence.

The radiocarbon content of some plant life has been linked to the distribution of smog in some areas. For example, the creosote bush in the Los Angeles area has been shown to have an effect on smog distribution that is more than fossil fuel combustion alone.

Health effects

Smog is a serious problem in many cities and continues to harm human health. Ground-level ozone, sulfur dioxide, nitrogen dioxide and carbon monoxide are especially harmful for senior citizens, children, and people with heart and lung conditions such as emphysema, bronchitis, and asthma. It can inflame breathing passages, decrease the lungs' working capacity, cause shortness of breath, pain when inhaling deeply, wheezing, and coughing. It can cause eye and nose irritation and it dries out the protective membranes of the nose and throat and interferes with the body's ability to fight infection, increasing susceptibility to illness. Hospital admissions and respiratory deaths often increase during periods when ozone levels are high.

Levels of unhealthy exposure

The U.S. EPA has developed an Air Quality Index to help explain air pollution levels to the general public. 8 hour average ozone concentrations of 85 to 104 ppbv are described as "Unhealthy for Sensitive Groups", 105 ppbv to 124 ppbv as "unhealthy" and 125 ppb to 404 ppb as "very unhealthy". The "very unhealthy" range for some other pollutants are: 355 ìg m"3 - 424 ìg m"3 for PM10; 15.5 ppm - 30.4ppm for CO and 0.65 ppm - 1.24 ppm for NO2.

Premature deaths due to cancer and respiratory disease

The Ontario Medical Association announced that smog is responsible for an estimated 9,500 premature deaths in the province each year. A 20-year American Cancer Society study found that cumulative exposure also increases the likelihood of premature death from a respiratory disease, implying the 8-hour standard may be insufficient.

Smog and the risk of certain birth defects

A study examining 806 women who had babies with birth defects between 1997 and 2006, and 849 women who had healthy babies, found that smog in the San Joaquin Valley area of California was linked to two types of neural tube defects: spina bifida (a condition involving, among other manifestations, certain malformations of the spinal column), and anencephaly (the underdevelopment or absence of part or all of the brain, which if not fatal usually results in profound impairment).

Smog and low birth weight

According to a study published in The Lancet, even a very small (5 ìg) change in PM 2.5 exposure was associated with an increase (18%) in risk of a low birth weight at delivery, and this relationship held even below the current accepted safe levels.

Areas affected

Smog can form in almost any climate where industries or cities release large amounts of air pollution, such as smoke or gases. However, it is worse during periods of warmer, sunnier weather when the upper air is warm enough to inhibit vertical circulation. It is especially prevalent in geologic basins encircled by hills or mountains. It often stays for an extended period of time over densely populated cities or urban areas, and can build up to dangerous levels.

ENVIRONMENTAL ISSUES IN DELHI

Environmental problems in Delhi,India, are a threat to the well-being of the city's and area's inhabitants as well as the flora and fauna. Delhi, the fifth-most populated metropolis in the world, is one of the most heavily polluted cities in

India, having for instance one of the country's highest volumes of particulate matter pollution. Overpopulation and the ensuing overuse of scarce resources such as water put heavy pressure on the environment. The city suffers from air pollution caused by road dust and industry, with comparatively smaller contributions from unclean engines in transportation, especially diesel-powered city buses and trucks, and 2-wheelers and 3-wheelers with two-stroke engines. Noise pollution comes mainly from motorcycle and automobile traffic. Water pollution and a lack of solid waste treatment facilities have caused serious damage to the river on whose banks Delhi grew, the Yamuna. Besides human and environmental damage, pollution has caused economic damage as well; Delhi may have lost the competition to host the 2014 Asian Games because of its poor environment.

Water pollution

Yamuna river

The river Yamuna, the reason for Delhi's existence, has suffered heavily from pollution. At its point of entry into Delhi, at Wazirabad, its dissolved oxygen (DO) content is 7.5 milligrammes per litre. At its point of exit from city limits, the DO level is only 1.3 mg/l. Similarly, coliform counts jump from 8,500 per 100 ml at entry to 329,312/100ml at exit (for DO 5 mg/litre is the norm and for coliforms 500/100ml). In 2007, roughly half of all the city's raw sewage went straight into the river. 55% of the city's 15 million people are connected to the city's sewer system and its treatment plants, but because of corrosion and clogging in the system many of the treatment plants do not run at full capacity. Waste from 1,500 unplanned neighborhoods runs straight into the river.

The Supreme Court of India took up the issue in 1994 after reports in the press, and since 2001 is actively monitoring the river and the city's efforts to clean it; in 2011, the national government announced a Rs 1,357 crore drain interceptor plan (all waste water is to be cleaned before it reaches the river) that would clean up the river by 2014.

Water sources

Underground hydrological resources are a substantial supplemental source of water in Delhi, especially in the affluent sections of the city. In the residential plots called 'farmhouses' almost every household draws from this resource. Though water-storing rocks, i.e. aquifers, are renewed as surface rain-water percolates down, they are not inexhaustible. Delhi's aquifers stand in danger of depletion on account of excessive use. Furthermore, rampant construction activity has contaminated them with cement, paints, varnishes and other construction materials; leaky, poorly constructed and maintained sewage lines have added to the contamination.

This is an irremediable loss, as aquifers, once polluted, cannot be decontaminated; they have no exposure to air and sunlight or to micro-organisms which clear-up chemical or biological pollutants.

Contributing further to underground water degradation are Delhi's mushrooming landfill sites. Waste material leeches underground, contaminating aquifers. Besides, land-fill sites degrade land. Delhi has twenty-five landfill sites, and more are planned.

Loss of flora and fauna

There is significant dispute over the extent of the city's green cover. City authorities claimed in 2008 that the green cover had increased from 26 km2 to 300 km2; moreover, the Delhi Forest Act stipulated that for every felled tree ten saplings need to be planted. Critics point out that the data as well as the meaning of "green cover" are unclear. The actual increase may be only half of what was claimed, and there are estimates that some 100,000 trees had been cut in Delhi, due in part to the construction of the Delhi Metroand the Delhi Bus Rapid Transit System.

Air pollution

Air pollution in Delhi is caused mainly by industry and vehicular traffic. As many as 10,000 people a year may die prematurely in Delhi as a result of air pollution. The 1997 White Paper sponsored by the Ministry of Environment and Forests already proposed various measures to bring down pollution caused by traffic, including smoothing the flow of traffic with parking regulations and bringing down total traffic by mandatory limits on driving. City authorities claim to have had some success in bringing down air pollution; for instance, during the bidding process for the 2014 Asian Games, the city's organizing committee had claimed that "pollution levels had come down drastically in Delhi with the arrival of Metro rail as well as all public transport vehicle being run compulsorily on compressed natural gas."

For traffic related sources, growth in vehicle numbers and mileage seems to outpace efforts to reduce emissions. Contrary to popular belief, most of the air pollution in Delhi is not due to vehicular traffic. Main contributors to particulate matter in the PM10 range, as a recent study shows, are road dust (50%) and industry (23%)—vehicles accounted for only 7%. Among industrial contributors, power plants within Delhi city limits were the main culprits.

Proposed solutions

The Delhi Development Authority (DDA) is charged with providing "lung spaces." Of the city's 44777 hectares, 8422 hectares are reserved for "the Greens", of which the DDA manages more than 5050 hectares. There is a policy for afforestation, atmospheric pollution, bio-medical waste, domestic refuse, and water and sewage treatment. Additionally, there are action plans to encourage public participation in environmental problems.

Given the continued growth of the city and its population, problems are tackled only with difficulty—for instance, the Yamuna clean-up projects spent $500 million between 1993 and 2005, yet the river's pollution actually doubled during this same period.

Environmentalists have criticised the Delhi government for not doing enough to curb air pollution and to inform people about air quality issues. Most of Delhi's residents are unaware of alarming levels of air pollution in the city and the health risks associated with it. Since the mid-1990s, Delhi has undertaken some measures to curb air pollution - Delhi has the third highest quantity of trees among Indian cities and the Delhi Transport Corporation operates the world's largest fleet of environmentally friendly compressed natural gas (CNG) buses. In 1996, the Centre for Science and Environment (CSE) started a public interest litigation in the Supreme Court of India that ordered the conversion of Delhi's fleet of buses and taxis to run on CNG and banned the use of leaded petrol in 1998. In 2003, Delhi won the United States Department of Energy's first 'Clean Cities International Partner of the Year' award for its "bold efforts to curb air pollution and support alternative fuel initiatives". The Delhi Metro has also been credited for significantly reducing air pollutants in the city.

However, according several authors, most of these gains have been lost, especially due to stubble burning, rise in market share of diesel cars and a considerable decline in bus ridership. According to CSE and System of Air Quality Weather Forecasting and Research (SAFAR), burning of agricultural waste in nearby Punjab, Haryana and Uttar Pradesh regions results in severe intensification of smog over Delhi. The state government of adjoining Uttar Pradesh is considering imposing a ban on crop burning to reduce pollution in Delhi NCR and an environmental panel has appealed to India's Supreme Court to impose a 30% cess on diesel cars.

United Kingdom

London

In 1306, concerns over air pollution were sufficient for Edward I to (briefly) ban coal fires in London. In 1661, John Evelyn's Fumifugium suggested burning fragrant wood instead of mineral coal, which he believed would reduce coughing. The Ballad of Gresham College the same year describes how the smoke "does our lungs and spirits choke, Our hanging spoil, and rust our iron."

Severe episodes of smog continued in the 19th and 20th centuries, mainly in the winter, and were nicknamed "pea-soupers". The Great Smog of 1952 darkened the streets of London and killed approximately 4,000 people in the short time of 4 days (a further 8,000 died from its effects in the following weeks and months). Initially a flu epidemic was blamed for the loss of life.

In 1956 the Clean Air Act started legally enforcing smokeless zones in the capital. There were areas where no soft coal was allowed to be burned in homes or in businesses, only coke, which produces no smoke. Because of the smokeless zones, reduced levels of sooty particulates made the intense and persistent London smog a thing of the past.

It was after this that the great clean-up of London began. One by one, historical buildings which, during the previous two centuries had gradually completely blackened externally, had their stone facades cleaned and restored to their original appearance. Victorian buildings whose appearance changed dramatically after cleaning included the British Museum of Natural History. A more recent example was the Palace of Westminster, which was cleaned in the 1980s. A notable exception to the restoration trend was 10 Downing Street, whose bricks upon cleaning in the late 1950s proved to be naturally yellow; the smog-derived black colour of the façade was considered so iconic that the bricks were painted black to preserve the image. Smog caused by traffic pollution, however, does still occur in modern London.

Other areas

Other areas of the United Kingdom were affected by smog, especially heavily industrialised areas.

The cities of Glasgow and Edinburgh, in Scotland, suffered smoke-laden fogs in 1909. Des Voeux, commonly credited with creating the "smog" moniker, presented a paper in 1911 to the Manchester Conference of the Smoke Abatement League of Great Britain about the fogs and resulting deaths. One Birmingham resident described near black-out conditions in the 1900s before the Clean Air Act, with visibility so poor that cyclists had to dismount and walk in order to stay on the road.

Mexico City, Mexico

Due to its location in a highland "bowl", cold air sinks down onto the urban area of Mexico City, trapping industrial and vehicle pollution underneath, and turning it into the most infamously smog-plagued city of Latin America. Within one generation, the city has changed from being known for some of the cleanest air of the world into one with some of the worst pollution, with pollutants like nitrogen dioxide being double or even triple international standards.

Santiago, Chile

Similar to Mexico City, the air pollution of Santiago valley, located between the Andes and the Chilean Coast Range, turn it into the most infamously smog-plagued city of South America. Other aggravates of the situation reside in its high latitude (31 degrees South) and dry weather during most of the year.

Tehran, Iran

In December 2005, schools and public offices had to close in Tehran, Iran and 1600 people were taken to hospital, in a severe smog blamed largely on unfiltered car exhaust.

United States

Smog was brought to the attention of the general US public in 1933 with the publication of the book "Stop That Smoke", by Henry Obermeyer, a New York public utility official, in which he pointed out the effect on human life and even the destruction of 3,000 acres (12 km^2) of a farmer's spinach crop. Since then, the United States Environmental Protection Agency has designated over 300 U.S. counties to be non-attainment areas for one or more pollutants tracked as part of the National Ambient Air Quality Standards. These areas are largely clustered around large metropolitan areas, with the largest contiguous non-attainment zones in California and the Northeast. Various U.S. and Canadian government agencies collaborate to produce real-time air quality maps and forecasts.

Los Angeles and the San Joaquin Valley

Because of their locations in low basins surrounded by mountains, Los Angeles and the San Joaquin Valley are notorious for their smog. The millions of vehicles in these regions combined with the additional effects of the San Francisco Bay and Los Angeles/Long Beach port complexes frequently contribute to further air pollution. While strict regulations by numerous California government agencies overseeing this problem have decreased the number of Stage 1 smog alerts from several hundred annually to just a few, these geologically predisposed entrapment zones collect pollution levels from cars, trucks and fixed sources which still exceeds health standards and is a pressing issue for the more than 25 million people who live there.

Major incidents in the US

- 1943, July 26, Los Angeles, California: A smog so sudden and severe that "Los Angeles residents believe the Japanese are attacking them with chemical warfare."
- 1948, October 30–31, Donora, Pennsylvania: 20 died, 600 hospitalized, thousands more stricken. Lawsuits were not settled until 1951.

Ulaanbaatar, Mongolia

In the late 1990s, massive immigration to Ulaanbaatar from the countryside began. An estimated 150,000 households, mainly living in traditional Mongolian gers on the outskirts of Ulaanbaatar, burn wood and coal (some poor families burn even car tires and trash) to heat themselves during the harsh winter, which lasts from October to April, since these outskirts are not connected to the city's central heating system. A temporary solution to decrease smog was proposed in the form of stoves with improved efficiency, although with no visible results. Coal-fired ger stoves release high levels of ash and other particulate matter (PM). When inhaled, these particles can settle in the lungs and respiratory tract and cause health problems. At two to 10 times above Mongolian and international air quality standards, Ulaanbaatar's PM rates are among the worst in the world, according to a December 2009 World Bank report. The Asian Development Bank (ADB) estimates that health costs related to this air pollution account for as much as 4 percent of Mongolia's GDP.

Southeast Asia

Smog is a regular problem in Southeast Asia caused by land and forest fires in Indonesia, especially Sumatra and Kalimantan, although the term haze is preferred in describing the problem. Farmers and plantation owners are usually responsible for the fires, which they use to clear tracts of land for further plantings. Those fires mainly affect Brunei, Indonesia, Philippines, Malaysia, Singapore and Thailand, and occasionally Guam and Saipan. The economic losses of the fires in 1997 have been estimated at more than US$9 billion. This includes damages in agriculture production, destruction of forest lands, health, transportation, tourism, and other economic endeavours. Not included are social, environmental, and psychological problems and long-term health effects. The second-latest bout of haze to occur in Malaysia, Singapore and the Malacca Straits is in October 2006, and was caused by smoke from fires in Indonesia being blown across the Straits of Malacca by south-westerly winds. A similar haze has occurred in June 2013, with the PSI setting a new record in Singapore on June 21 at 12pm with a reading of 401, which is in the "Hazardous" range.

The Association of Southeast Asian Nations (ASEAN) reacted. In 2002, the Agreement on Transboundary Haze Pollution was signed between all ASEAN nations. ASEAN formed a Regional Haze Action Plan (RHAP) and established a co-ordination and support unit (CSU). RHAP, with the help of Canada, established a monitoring and warning system for forest/vegetation fires and implemented a Fire Danger Rating System (FDRS). The Malaysian Meteorological Department (MMD) has issued a daily rating of fire danger since September 2003. Indonesia has been ineffective at enforcing legal policies on errant farmers.

Pollution index

The severity of smog is often measured using automated optical instruments such as Nephelometers, as haze is associated with visibility and traffic control in ports. Haze however can also be an indication of poor air quality though this is often better reflected using accurate purpose built air indexes such as the American Air Quality Index, the Malaysian API (Air Pollution Index) and the Singaporean Pollutant Standards Index. In hazy conditions, it is likely that the index will report the suspended particulate level. The disclosure of the responsible pollutant is mandated in some jurisdictions.

The Malaysian API does not have a capped value; hence its most hazardous readings can go above 500. Above 500, a state of emergency is declared in the affected area. Usually, this means that non-essential government services are suspended, and all ports in the affected area are closed. There may also be prohibitions on private sector commercial and industrial activities in the affected area excluding the food sector. So far, state of emergency rulings due to hazardous API levels were applied to the Malaysian towns of Port Klang, Kuala Selangor and the state of Sarawak during the 2005 Malaysian haze and the 1997 Southeast Asian haze.

Cultural references

- The London "pea-soupers" earned the capital the nickname of "The Smoke". Similarly, Edinburgh was known as "Auld Reekie". The smogs feature in many London novels as a motif indicating hidden danger or a mystery, perhaps most overtly in Margery Allingham's The Tiger in the Smoke (1952), but also in Dickens's Bleak House (1852) and T.S. Eliot's "The Love Song of J. Alfred Prufrock."
- The 1970 made-for-TV movie A Clear and Present Danger was one of the first American television network entertainment programs to warn about

the problem of smog and air pollution, as it dramatized a man's efforts toward clean air after emphysema killed his friend.

- The history of smog in LA is detailed in Smogtown by Chip Jacobs and William J. Kelly.
- The London "pea-soupers" earned the capital the nickname of "The Smoke". Similarly, Edinburgh was known as "Auld Reekie". The smogs feature in many London novels as a motif indicating hidden danger or a mystery, perhaps most overtly in Margery Allingham's The Tiger in the Smoke (1952), but also in Dickens's Bleak House (1852) and T.S. Eliot's "The Love Song of J. Alfred Prufrock."

5

DRUG ACTION

The action of drugs on the human body is called pharmacodynamics, and what the body does with the drug is called pharmacokinetics. The drugs that enter the human tend to stimulate certain receptors, ion channels, act on enzymes or transporter proteins. As a result, they cause the human body to react in a specific way.

There are 2 different types of drugs:

- Agonists - they stimulate and activate the receptors
- Antagonists - they stop the agonists from stimulating the receptors

Once the receptors are activated, they either trigger a particular response directly on the body, or they trigger the release of hormones and/or other endogenous drugs in the body to stimulate a particular response.

SHORT NOTE ON RECEPTORS

Drugs interact with receptors by bonding at specific binding sites. Most receptors are made up of proteins, and the drugs can therefore interact with the amino acids to change the conformation of the receptor proteins.

These interactions are very basic, just like that of other chemical bondings:

Ionic bonds

Mainly occur through attractions between opposite charges; for example, between protonated amino (on salbutamol) or quaternary ammonium (e.g. acetylcholine), and the dissociated carboxylic acid group. Similarly, the dissociated carboxylic acid group on the drug can bind with amino groups on the receptor.

This type of bond is very strong, and varies with the inverse of the distance between the atoms so that it can act over large distances.

Cation-ð interactions can also be classified as ionic bonding. This type of interaction occurs when a cation, e.g. acetylcholine, interacts with the negative ð bonds on an aromatic group of the receptor.

Ion-dipole and dipole-dipole bonds have similar interactions, but are more complicated and are weaker than ionic bonds.

Hydrogen bonds

There is a small but significant attraction between hydrogen atoms and polar functional groups (e.g. the hydroxyl [-OH] group). These so-called hydrogen bonds only act over short distances, and are dependent on the correct alignment between functional groups.

Receptors are located on all cells in the body. The same receptor can be located on different organs, and even on different types of tissues. There are also different subtypes of receptor which elicit different effects in response to the same agonist. For example, there are two types of histamine receptor: H_1 and H_2. Activation of the H_1 subtype receptor causes contraction of smooth muscle, whereas activation of the H_2 receptor stimulates gastric secretion.

It is this phenomenon that gives rise to drug specificity. Of course, drugs do not only act on receptors: they also act on ion channels, enzymes, and cell transporter proteins. these hydrogen bonds are so powerful than the ionic bonds.

How shape of drug molecules affect drug action

When talking about the shape of molecules, biochemists are mainly concerned with the three-dimensional conformation of drug molecules. There are many isomers of a particular drug, and each one will have its own effects. Differences in isomer affect not only what the drug activates, but also changes the potency of each drug.

Potency

Potency is a measure of how much a drug is required in order to produce a particular effect. Therefore, only a small dosage of a high potency drug is required to induce a large response. The other terms used to measure the ability of a drug to trigger a response are:

- Intrinsic Activity which defines:
- Agonists as having Intrinsic Activity = 1
- Antagonists as having Intrinsic Activity = 0
- Partial Agonist as having Intrinsic Activity between 0 and 1

- Intrinsic Efficacy also measures the different activated state of receptors, and the ability for a drug to cause maximum response without having to bind to all the receptors.

The specificity of drugs

Drug companies invest significant effort in designing drugs that interact specifically with particular receptors, since non-specific drugs can cause more side effects.

An example is the endogenous drug acetylcholine (ACh). ACh is used by the parasympathetic nervous system to activate muscarinic receptors and by the neuromuscular system to activate nicotinic receptors. However, the compounds muscarine and nicotine can each preferentially interact one of the two receptor types, allowing them to activate only one of the two systems where ACh itself would activate both.

AFFINITY

The specificity of drugs cannot be talked about without mentioning the affinity of the drugs. The affinity is a measure of how tightly a drug binds to the receptor. If the drug does not bind well, then the action of the drug will be shorter and the chance of binding will also be less. This can be measured numerically by using the dissociation constant KD. The value of KD is the same as the concentration of drug when 50% of receptors are occupied.

But the value of KD is also affected by the conformation, bonding and size of the drug and the receptor. The higher the KD the lower the affinity of the drug.

Pharmacodynamics

Pharmacodynamics (PD) is the study of the biochemical and physiologic effects of drugs (especially pharmaceutical drugs). The effects can include those manifested within animals(including humans), microorganisms, or combinations of organisms (for example, infection). Pharmacodynamics is the study of how a drug affects an organism, whereas pharmacokinetics is the study of how the organism affects the drug. Both together influence dosing, benefit, and adverse effects. Pharmacodynamics is sometimes abbreviated as PD and pharmacokinetics as PK, especially in combined reference (for example, when speaking of PK/PD models).

Pharmacodynamics places particular emphasis on dose–response relationships, that is, the relationships between drug concentration and effect.

Effects on the body

The majority of drugs either

a. mimic or inhibit normal physiological/biochemical processes or inhibit pathological processes in animals or

b. inhibit vital processes of endo- or ectoparasites and microbial organisms.

There are 7 main drug actions:

- Stimulating action through direct receptor agonism and downstream effects
- Depressing action through direct receptor agonism and downstream effects (ex.: inverse agonist)
- Blocking/antagonizing action (as with silent antagonists), the drug binds the receptor but does not activate it
- Stabilizing action, the drug seems to act neither as a stimulant or as a depressant (ex.: some drugs possess receptor activity that allows them to stabilize general receptor activation, like buprenorphine in opioid dependent individuals or aripiprazole in schizophrenia, all depending on the dose and the recipient)
- Exchanging/replacing substances or accumulating them to form a reserve (ex.: glycogen storage)
- Direct beneficial chemical reaction as in free radical scavenging
- Direct harmful chemical reaction which might result in damage or destruction of the cells, through induced toxic or lethal damage (cytotoxicity or irritation)

Desired activity

The desired activity of a drug is mainly due to successful targeting of one of the following:

- Cellular membrane disruption
- Chemical reaction with downstream effects
- Interaction with enzyme proteins
- Interaction with structural proteins
- Interaction with carrier proteins
- Interaction with ion channels

- Ligand binding to receptors:
- Hormone receptors
- Neuromodulator receptors
- Neurotransmitter receptors

General anesthetics were once thought to work by disordering the neural membranes, thereby altering the Na influx. Antacids and chelating agents combine chemically in the body. Enzyme-substrate binding is a way to alter the production or metabolism of key endogenous chemicals, for example aspirin irreversibly inhibits the enzyme prostaglandin synthetase (cyclooxygenase) thereby preventing inflammatory response. Colchicine, a drug for gout, interferes with the function of the structural protein tubulin, while Digitalis, a drug still used in heart failure, inhibits the activity of the carrier molecule, Na-K-ATPase pump. The widest class of drugs act as ligands which bind to receptors which determine cellular effects. Upon drug binding, receptors can elicit their normal action (agonist), blocked action (antagonist), or even action opposite to normal (inverse agonist).

In principle, a pharmacologist would aim for a target plasma concentration of the drug for a desired level of response. In reality, there are many factors affecting this goal. Pharmacokinetic factors determine peak concentrations, and concentrations cannot be maintained with absolute consistency because of metabolic breakdown and excretory clearance. Genetic factors may exist which would alter metabolism or drug action itself, and a patient's immediate status may also affect indicated dosage.

Undesirable effects

Undesirable effects of a drug include:

- Increased probability of cell mutation (carcinogenic activity)
- A multitude of simultaneous assorted actions which may be deleterious
- Interaction (additive, multiplicative, or metabolic)
- Induced physiological damage, or abnormal chronic conditions

THERAPEUTIC WINDOW

The therapeutic window is the amount of a medication between the amount that gives an effect (effective dose) and the amount that gives more adverse effects than desired effects. For instance, medication with a small pharmaceutical window must be administered with care and control, e.g. by frequently measuring blood concentration of the drug, since it easily loses effects or gives adverse effects.

Duration of action

The duration of action of a drug is the length of time that particular drug is effective. Duration of action is a function of several parameters including plasma half-life, the time to equilibrate between plasma and target compartments, and the off rate of the drug from its biological target.

Receptor binding and effect

The binding of ligands (drug) to receptors is governed by the law of mass action which relates the large-scale status to the rate of numerous molecular processes. The rates of formation and un-formation can be used to determine the equilibrium concentration of bound receptors. The equilibrium dissociation constant is defined by: where L=ligand, R=receptor, square brackets [] denote concentration. The fraction of bound receptors is Semi-log plots of two agonists with different Kd. The blue curve represents the ligand with greater potency.

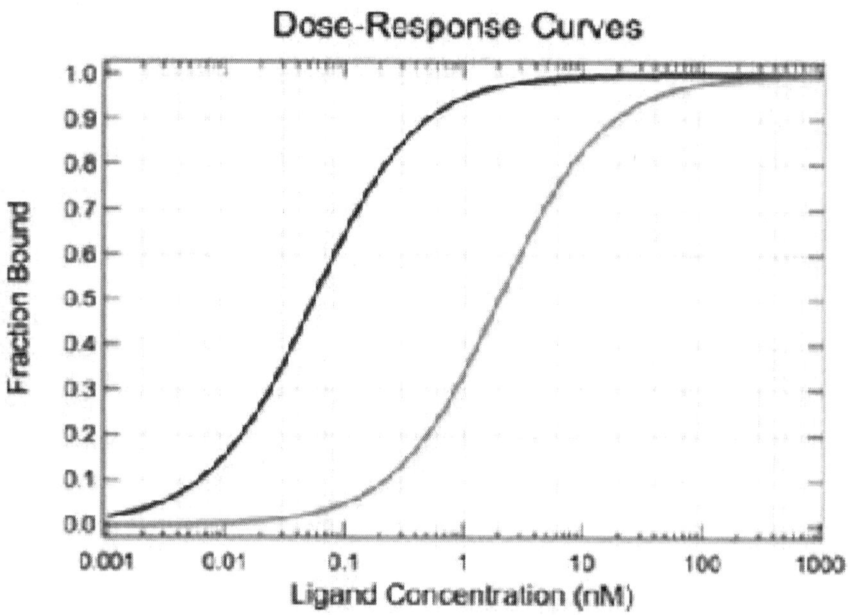

This expression is one way to consider the effect of a drug, in which the response is related to the fraction of bound receptors (see: Hill equation). The fraction of bound receptors is known as occupancy. The relationship between occupancy and pharmacological response is usually non-linear. This explains the so-called receptor reserve phenomenon i.e. the concentration producing 50% occupancy is typically higher than the concentration producing 50% of maximum response. More precisely, receptor reserve refers to a phenomenon whereby stimulation of only a fraction of the whole receptor population apparently elicits the maximal effect achievable in a particular tissue.

The simplest interpretation of receptor reserve is that it is a model that states there are excess receptors on the cell surface than what is necessary for full effect. Taking a more sophisticated approach, receptor reserve is an integrative measure of the response-inducing capacity of an agonist (in some receptor models it is termed intrinsic efficacy or intrinsic activity) and of the signal amplification capacity of the corresponding receptor (and its downstream signaling pathways). Thus, the existence (and magnitude) of receptor reserve depends on the agonist (efficacy), tissue (signal amplification ability) and measured effect (pathways activated to cause signal amplification). As receptor reserve is very sensitive to agonist's intrinsic efficacy, it is usually defined only for full (high-efficacy) agonists.

Often the response is determined as a function of log[L] to consider many orders of magnitude of concentration. However, there is no biological or physical theory which relates effects to the log of concentration. It is just convenient for graphing purposes. It is useful to note that 50% of the receptors are bound when [L]=Kd .

The graph shown represents the conc-response for two hypothetical receptor agonists, plotted in a semi-log fashion. The curve toward the left represents a higher potency (potency arrow does not indicate direction of increase) since lower concentrations are needed for a given response. The effect increases as a function of concentration.

MULTICELLULAR PHARMACODYNAMICS

The concept of pharmacodynamics has been expanded to include Multicellular Pharmacodynamics (MCPD). MCPD is the study of the static and dynamic properties and relationships between a set of drugs and a dynamic and diverse multicellular four-dimensional organization. It is the study of the workings of a drug on a minimal multicellular system (mMCS), both in vivo and in silico. Networked Multicellular Pharmacodynamics (Net-MCPD) further extends the concept of MCPD to model regulatory genomic networks together with signal transduction pathways, as part of a complex of interacting components in the cell.

Toxicodynamics

Pharmacokinetics and pharmacodynamics are termed toxicokinetics and toxicodynamics in the field of ecotoxicology. Here, the focus is on toxic effects on a wide range of organisms. The corresponding models are called toxicokinetic-toxicodynamic models.

Mechanism of action

In pharmacology, the term mechanism of action (MOA) refers to the specific biochemical interaction through which a drug substance produces its pharmacological effect. A mechanism of action usually includes mention of the specific molecular targets to which the drug binds, such as an enzyme or receptor. Receptor sites have specific affinities for drugs based on the chemical structure of the drug, as well as the specific action that occurs there. Drugs that do not bind to receptors produce their corresponding therapeutic effect by simply interacting with chemical or physical properties in the body. Common examples of drugs that work in this way are antacids and laxatives.

In comparison, a mode of action (MoA) describes functional or anatomical changes, at the cellular level, resulting from the exposure of a living organism to a substance.

Why mechanism of action is important

Elucidating the mechanism of action of novel drugs and medications is important for several reasons:

- In the case of anti-infective drug development, the information permits anticipation of problems relating to clinical safety. Drugs disrupting the cytoplasmic membrane or electron transport chain, for example, are more likely to cause toxicity problems than those targeting components of the cell wall (peptidoglycan or â-glucans) or 70S ribosome, structures which are absent in human cells.
- By knowing the interaction between a certain site of a drug and a receptor, other drugs can be formulated in a way that replicates this interaction, thus producing the same therapeutic effects. Indeed, this method is used to create new drugs.
- It can help identify which patients are most likely to respond to treatment. Because the breast cancer medication trastuzumab is known to target protein HER2, for example, tumors can be screened for the presence of this molecule to determine whether or not the patient will benefit from trastuzumab therapy.
- It can enable better dosing because the drug's effects on the target pathway can be monitored in the patient. Statin dosage, for example, is usually determined by measuring the patient's blood cholesterol levels.
- It allows drugs to be combined in such a way that the likelihood of drug

resistance emerging is reduced. By knowing what cellular structure an anti-infective or anticancer drug acts upon, it is possible to administer a cocktail that inhibits multiple targets simultaneously, thereby reducing the risk that a single mutation in microbial or tumor DNA will lead to drug resistance and treatment failure.

- It may allow other indications for the drug to be identified. Discovery that sildenafil inhibits phosphodiesterase-5 (PDE-5) proteins, for example, enabled this drug to be repurposedfor pulmonary arterial hypertension treatment, since PDE-5 is expressed in pulmonary hypertensive lungs.

HOW MECHANISM OF ACTION IS DETERMINED

Microscopy-based methods

Bioactive compounds induce phenotypic changes in target cells, changes that are observable by microscopy, and which can give insight into the mechanism of action of the compound. With antibacterial agents, for example, the conversion of target cells to spheroplasts can be an indication that peptidoglycan synthesis is being targeted, and filamentation of target cells can be an indication that FtsZ or DNA is being targeted. In the case of anticancer agents, bleb formation can be an indication that the compound is disrupting the plasma membrane. A current limitation of this approach is the time required to manually generate and interpret data, but advances in automated microscopy and image analysis software may help resolve this.

Direct biochemical methods

Direct biochemical methods include methods in which a protein or a small molecule, such as a drug candidate, is labeled and is traced throughout the body. This proves to be the most direct approach to find target protein that will bind to small targets of interest, such as a basic representation of a drug outline, in order to identify the pharmacophore of the drug. Due to the physical interactions between the labeled molecule and a protein, biochemical methods can be used to determine the toxicity, efficacy, and the mechanism of action of the drug.

Computation inference methods

Typically, computation inference methods are primarily used to predict protein targets for small molecule drugs based on computer based pattern recognition. However, this method could also be used for finding new targets for existing or newly developed drugs. By identifying the pharmacophore of the drug molecule, the profiling method of pattern recognition can be carried out where a new target is identified. This provides an insight at a possible mechanism of action, as it is known what certain functional components of the drug are responsible for interacting with a certain area on a protein, thus, leading to a therapeutic effect.

Drugs with known mechanisms of action

There are many drugs in which the mechanism of action is known. One example is aspirin.

Aspirin

The mechanism of action of aspirin involves irreversible inhibition of the enzyme cyclooxygenase; therefore suppressing the production of prostaglandins and thromboxanes, thus reducing pain and inflammation. This mechanism of action is specific to aspirin, and is not constant for all nonsteroidal anti-inflammatory drugs (NSAIDs). Rather, aspirin is the only NSAID that irreversibly inhibits COX-1.

Drug interaction

Certain medicines can interact pharmacologically and affect the activity of other medicines.

A drug interaction, as the name suggests, is when two or more drugs interact when administered together. This result can be synergistic (the desired effect is increased) or antagonistic (the desired effect is decreased). Alternatively, a new effect can be produced which neither drug produces on its own. Interactions may also exist between drugs and foods. An example would be people taking antidepressant drugs such as monoamine oxidase inhibitors (MAOIs) should not take food containing tyramine as hypertensive crisis may occur. There are also possible interaction between drugs and medicinal plants or herbs, (herb-drug interactions). These interactions may occur out of accidental misuse, or due to lack of knowledge about the active ingredients involved in the relevant substances.

It is therefore easy to see the importance of these pharmacological interactions in the practice of medicine. If a patient is taking two drugs and one of them increases the effect of the other, it is possible that an overdose may occur. The interaction of the two drugs may also increase the risk that side effects will occur. On the other hand, if the action of a drug is reduced, it may cease to have any therapeutic use because of under dosage. Notwithstanding the above, on occasion these interactions may be sought in order to obtain an improved therapeutic effect. Examples of this include the use of codeine with paracetamol to increase their analgesic effect or the combination of clavulanic acid with amoxicillin in order to overcome bacterial resistance to the antibiotic. It should also be remembered that there are interactions that, from a theoretical standpoint, may occur but in clinical practice have no important repercussions.

The pharmaceutical interactions that are of special interest to the practice of medicine are primarily those that have negative effects for an organism. The risk that a pharmacological interaction will appear increases as a function of the number of drugs administered to a patient at the same time. Over a third (36%) of older adults in the U.S. regularly use 5 or more medications or supplements and 15% are potentially at risk for a major drug-drug interaction. Both the use of medications and subsequent adverse drug interactions have increased significantly between 2005-2011.

It is possible that an interaction will occur between a drug and another substance present in the organism (*i.e.* foods or alcohol). Or in certain specific situations a drug may even react with itself, such as occurs with dehydration. In other situations, the interaction does not involve any effect on the drug. In certain cases, the presence of a drug in an individual's blood may affect certain types of laboratory analysis (analytical interference).

It is also possible for interactions to occur outside an organism before administration of the drugs has taken place. This can occur when two drugs are mixed, for example, in a saline solution prior to intravenous injection. Some classic examples of this type of interaction include that thiopentone and suxamethonium should not be placed in the same syringe and same is true for benzylpenicillin and heparin. These situations will all be discussed under the same heading due to their conceptual similarity.

Drug interactions may be the result of various processes. These processes may include alterations in the pharmacokinetics of the drug, such as alterations in the absorption, distribution, metabolism, and excretion (ADME) of a drug. Alternatively, drug interactions may be the result of the pharmacodynamic properties of the drug, e.g. the co-administration of a receptor antagonist and an agonist for the same receptor.

SYNERGY AND ANTAGONISM

When the interaction causes an increase in the effects of one or both of the drugs the interaction is called a synergistic effect. An "additive synergy" occurs when the final effect is equal to the sum of the effects of the two drugs (Although some authors argue that this is not true synergy). When the final effect is much greater than the sum of the two effects this is called enhanced synergy. This concept is recognized by the majority of authors, although other authors only refer to synergy when there is an enhanced effect. These authors use the term "additive effect" for additive synergy and they reserve use of the term "synergistic effect" for enhanced synergy. The opposite effect to synergy is termed antagonism. Two drugs are antagonistic when their interaction causes a decrease in the effects of one or both of the drugs.

Both synergy and antagonism can both occur during different phases of the interaction of a drug with an organism, with each effect having a different name. For example, when the synergy occurs at a cellular receptor level this is termed agonism, and the substances involved are termed agonists. On the other hand, in the case of antagonism the substances involved are known as inverse agonists. The different responses of a receptor to the action of a drug has resulted in a number of classifications, which use terms such as "partial agonist", "competitive agonist" etc. These concepts have fundamental applications in the pharmacodynamics of these interactions. The proliferation of existing classifications at this level, along with the fact that the exact reaction mechanisms for many drugs are not well understood means that it is almost impossible to offer a clear classification for these concepts. It is even likely that many authors would misapply any given classification.

Underlying Factors

It is possible to take advantage of positive drug interactions. However, the negative interactions are usually of more interest because of their pathological significance and also because they are often unexpected and may even go undiagnosed. By studying the conditions that favour the appearance of interactions it should be possible to prevent them or at least diagnose them in time. The factors or conditions that predispose or favor the appearance of interactions include:

- Old age: factors relating to how human physiology changes with age may affect the interaction of drugs. For example, liver metabolism, kidney function, nerve transmission or the functioning of bone marrow all decrease with age. In addition, in old age there is a sensory decrease that increases the chances of errors being made in the administration of drugs.
- Polypharmacy: The more drugs a patient takes the more likely it will be that some of them will interact.
- Genetic factors: Genes synthesize enzymes that metabolize drugs. Some races have genotypic variations that could decrease or increase the activity of these enzymes. The consequence of this would, on occasions, be a greater predisposition towards drug interactions and therefore a greater predisposition for adverse effects to occur. This is seen in genotype

variations in the isozymes of cytochrome P450.

- Hepatic or renal diseases: The blood concentrations of drugs that are metabolized in the liver and / or eliminated by the kidneys may be altered if either of these organs is not functioning correctly. If this is the case an increase in blood concentration is normally seen.
- Serious diseases that could worsen if the dose of the medicine is reduced.
- Drug dependent factors:
- Narrow therapeutic index: Where the difference between the effective dose and the toxic dose is small. The drug digoxin is an example of this type of drug.
- Steep dose-response curve: Small changes in the dosage of a drug produce large changes in the drug's concentration in the patient's blood plasma.
- Saturable hepatic metabolism: In addition to dose effects the capacity to metabolize the drug is greatly decreased

Analytical interference

The detection of laboratory parameters is based on physicochemical reactions between the substance being measured and reagents designed for this purpose. These reactions can be altered by the presence of drugs giving rise to an over estimation or an underestimation of the real results. Levels of cholesterol and other blood lipids can be overestimated as a consequence of the presence in the blood of some psychotropic drugs. These overestimates should not be confused with the action of other drugs that actually increase blood cholesterol levels due to an interaction with its metabolism. Most experts consider that these are not true interactions, so they will not be dealt with further in this discussion.

These chemical reactions are also known as pharmacological incompatibilities. The reactions occur when two or more drugs are mixed outside the body of the organism for the purpose of joint administration. Usually the interaction is antagonistic and it almost always affects both drugs. Examples of these types of interactions include the mixing of penicillins and aminoglycosides in the same serum bottle, which causes the formation of an insoluble precipitate, or the mixing of ciprofloxacin with furosemide. The interaction of some drugs with the transport medium can also be included here. This means that certain drugs cannot be administered in plastic bottles because they bind with the bottle's walls, reducing the drug's concentration in solution.

Many authors do not consider them to be interactions in the strictest sense of the word. An example is the database of the General Council of Official Pharmacists Colleges of Spain (Consejo General de Colegios Oficiales de Farmacéuticos de España), that does not include them among the 90,000 registered interactions.

Pharmacodynamic interactions

The change in an organism's response on administration of a drug is an important factor in pharmacodynamic interactions. These changes are extraordinarily difficult to classify given the wide variety of modes of action that exist and the fact that many drugs can cause their effect through a number of different mechanisms. This wide diversity also means that, in all but the most obvious cases, it is important to investigate and understand these mechanisms. The well-founded suspicion exists that there are more unknown interactions than known ones.

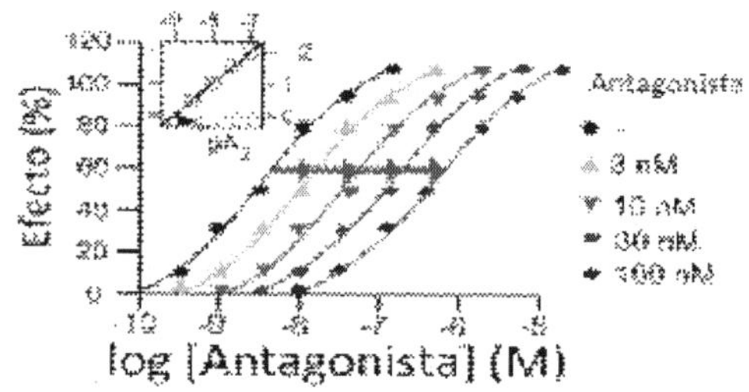

Effects of the competitive inhibition of an agonist by increases in the concentration of an antagonist. A drugs potency can be affected (the response curve shifted to the right) by the presence of an antagonistic interaction.pA2 known as the Schild representation, a mathematical model of the agonist:antagonist relationship or vice versa.

Pharmacokinetic interactions

Modifications in the effect of a drug are caused by differences in the absorption, transport, distribution, metabolization or excretion of one or both of the drugs compared with the expected behaviour of each drug when taken individually. These changes are basically modifications in the concentration of the drugs. In this respect two drugs can be homergic if they have the same effect in the organism and heterergic if their effects are different.

Absorption interactions

Changes in motility

Some drugs, such as the prokinetic agents increase the speed with which a substance passes through the intestines. If a drug is present in the digestive tract's absorption zone for less time its blood concentration will decrease. The opposite will occur with drugs that decrease intestinal motility.

- pH: Drugs can be present in either ionised or non-ionised form, depending on their pKa (pH at which the drug reaches equilibrium between its ionised and non-ionised form).The non-ionized forms of drugs are usually easier to absorb, because they will not be repelled by the lipidic bylayer of the cell, most of them can be absorbed by passive diffusion, unless they are too big or too polarized (like glucose or vancomicyn), in which case they may have or not specific and non specific transporters distributed on the entire intestine internal surface, that carries drugs inside the body. Obviously increasing the absorption of a drug will increase its bioavailability, so, changing the drug's state between ionized or not, can be useful or not for certain drugs.

Certain drugs require an acid stomach pH for absorption. Others require the basic pH of the intestines. Any modification in the pH could change this absorption. In the case of the antacids, an increase in pH can inhibit the absorption of other drugs such as zalcitabine (absorption can be decreased by 25%), tipranavir (25%) and amprenavir (up to 35%). However, this occurs less often than an increase in pH causes an increase in absorption. Such as occurs when cimetidine is taken with didanosine. In this case a gap of two to four hours between taking the two drugs is usually sufficient to avoid the interaction.

- Drug solubility: The absorption of some drugs can be drastically reduced if they are administered together with food with a high fat content. This is the case for oral anticoagulantsand avocado.
- Formation of non-absorbable complexes:
- Chelation: The presence of di- or trivalent cations can cause the chelation of certain drugs, making them harder to absorb. This interaction frequently occurs between drugs such as tetracycline or the fluoroquinolones and dairy products (due to the presence of Ca).
- Binding with proteins. Some drugs such as sucralfate binds to proteins, especially if they have a high bioavailability. For this reason its administration is contraindicated in enteral feeding.
- Finally, another possibility is that the drug is retained in the intestinal lumen forming large complexes that impede its absorption. This can occur with cholestyramine if it is associated with sulfamethoxazol, thyroxine, warfarin or digoxin.

- Acting on the P-glycoprotein of the enterocytes: This appears to be one of the mechanisms promoted by the consumption of grapefruit juice in increasing the bioavailability of various drugs, regardless of its demonstrated inhibitory activity on first pass metabolism.

TRANSPORT AND DISTRIBUTION INTERACTIONS

The main interaction mechanism is competition for plasma protein transport. In these cases the drug that arrives first binds with the plasma protein, leaving the other drug dissolved in the plasma, which modifies its concentration. The organism has mechanisms to counteract these situations (by, for example, increasing plasma clearance), which means that they are not usually clinically relevant. However, these situations should be taken into account if other associated problems are present such as when the method of excretion is affected.

Metabolism interactions

Many drug interactions are due to alterations in drug metabolism. Further, human drug-metabolizing enzymes are typically activated through the engagement of nuclear receptors. One notable system involved in metabolic drug interactions is the enzyme system comprising the cytochrome P450 oxidases.

CYP450

Cytochrome P450 is a very large family of haemoproteins (hemoproteins) that are characterized by their enzymatic activity and their role in the metabolism of a large number of drugs. Of the various families that are present in human beings the most interesting in this respect are the 1, 2 and 3, and the most important enzymes are CYP1A2, CYP2C9, CYP2C19, CYP2D6, CYP2E1 and CYP3A4. The majority of the enzymes are also involved in the metabolism of endogenous substances, such as steroids or sex hormones, which is also important should there be interference with these substances. As a result of these interactions the function of the enzymes can either be stimulated (enzyme induction) or inhibited (enzyme inhibition).

Enzymatic inhibition

If drug A is metabolized by a cytochrome P450 enzyme and drug B inhibits or decreases the enzyme's activity, then drug A will remain with high levels in the plasma for longer as its inactivation is slower. As a result, enzymatic inhibition will cause an increase in the drug's effect. This can cause a wide range of adverse reactions.

It is possible that this can occasionally lead to a paradoxical situation, where the enzymatic inhibition causes a decrease in the drug's effect: if the metabolism of drug A gives rise to product A2, which actually produces the effect of the drug. If the metabolism of drug A is inhibited by drug B the concentration of A2 that is present in the blood will decrease, as will the final effect of the drug.

Enzymatic induction

If drug A is metabolized by a cytochrome P450 enzyme and drug B induces or increases the enzyme's activity, then blood plasma concentrations of drug A will quickly fall as its inactivation will take place more rapidly. As a result, enzymatic induction will cause a decrease in the drug's effect.

As in the previous case it is possible to find paradoxical situations where an active metabolite causes the drug's effect. In this case the increase in active metabolite A2 (following the previous example) produces an increase in the drug's effect.

It can often occur that a patient is taking two drugs that are enzymatic inductors, one inductor and the other inhibitor or both inhibitors, which greatly complicates the control of an individual's medication and the avoidance of possible adverse reactions.

Enzyme CYP3A4 is the enzyme that the greatest number of drugs use as a substrate. Over 100 drugs depend on its metabolism for their activity and many others act on the enzyme as inductors or inhibitors.

Some foods also act as inductors or inhibitors of enzymatic activity. The following table shows the most common:

Any study of pharmacological interactions between particular medicines should also discuss the likely interactions of some medicinal plants. The effects caused by medicinal plants should be considered in the same way as those of medicines as their interaction with the organism gives rise to a pharmacological response. Other drugs can modify this response and also the plants can give rise to changes in the effects of other active ingredients. There is little data available regarding

1. False sense of security regarding medicinal plants. The interaction between a medicinal plant and a drug is usually overlooked due to a belief in the "safety of medicinal plants."
2. Variability of composition, both qualitative and quantitative. The composition of a plant-based drug is often subject to wide variations due to a number of factors such as seasonal differences in concentrations, soil type, climatic changes or the existence of different varieties or chemical

races within the same plant species that have variable compositions of the active ingredient. On occasions an interaction can be due to just one active ingredient, but this can be absent in some chemical varieties or it can be present in low concentrations, which will not cause an interaction. Counter interactions can even occur. This occurs, for instance, with ginseng, the Panax ginseng variety increases the Prothrombin time, while the Panax quinquefolius variety decreases it.

3. Absence of use in at-risk groups, such as hospitalized and polypharmacy patients, who tend to have the majority of drug interactions.
4. Limited consumption of medicinal plants has given rise to a lack of interest in this area.

They are usually included in the category of foods as they are usually taken as a tea or food supplement. However, medicinal plants are increasingly being taken in a manner more often associated with conventional medicines: pills, tablets, capsules, etc.

EXCRETION INTERACTIONS

Renal excretion

Only the free fraction of a drug that is dissolved in the blood plasma can be removed through the kidney. Therefore, drugs that are tightly bound to proteins are not available for renal excretion, as long as they are not metabolized when they may be eliminated as metabolites. Creatinine clearance is used as a measure of kidney functioning but it is only useful in cases where the drug is excreted in an unaltered form in the urine. The excretion of drugs from the kidney's nephrons has the same properties as that of any other organic solute: passive filtration, reabsorption and active secretion. In the latter phase the secretion of drugs is an active process that is subject to conditions relating to the saturability of the transported molecule and competition between substrates. Therefore, these are key sites where interactions between drugs could occur. Filtration depends on a number of factors including the pH of the urine, it having been shown that the drugs that act as weak bases are increasingly excreted as the pH of the urine becomes more acidic, and the inverse is true for weak acids. This mechanism is of great use when treating intoxications (by making the urine more acidic or more alkali) and it is also used by some drugs and herbal products to produce their interactive effect.

Bile excretion

Bile excretion is different from kidney excretion as it is always involves energy expenditure in active transport across the epithelium of the bile duct against a concentration gradient. This transport system can also be saturated if the plasma concentrations of the drug are high. Bile excretion of drugs mainly takes place where their molecular weight is greater than 300 and they contain both polar and lipophilic groups. The glucuronidation of the drug in the kidney also facilitates bile excretion. Substances with similar physicochemical properties can block the receptor, which is important in assessing interactions. A drug excreted in the bile duct can occasionally be reabsorbed by the intestines (in the entero-hepatic circuit), which can also lead to interactions with other drugs.

Epidemiology

Among US adults older than 55, 4% are taking medication and or supplements that put them at risk of a major drug interaction. Potential drug-drug interactions have increased over time and are more common in the low educated elderly even after controlling for age, sex, place of residence, and comorbidity.

6

DRUG ACTING ON CNS

Leaves of the coca plant (Erythroxylum novogranatense var. Novogranatense), from which cocaine , the only naturally occurring local anesthetic, is derived.

An anesthetic (American English) or anaesthetic (British English; see spelling differences) is a drug used to induce anesthesia - in other words, to result in a temporary loss of sensation or awareness. They may be divided into two broad classes: general anesthetics, which cause a reversible loss of consciousness, and local anesthetics, which cause a reversible loss of sensation for a limited region of the body without necessarily affecting consciousness.

A wide variety of drugs are used in modern anesthetic practice. Many are rarely used outside anesthesiology, but others are used commonly in various fields of healthcare. Combinations of anesthetics are sometimes used for their synergistic and additive therapeutic effects. Adverse effects, however, may also be increased. Anesthetics are distinct from analgesics, which block only sensation of painful stimuli.

LOCAL ANESTHETICS

Local anesthetic agents prevent transmission of nerve impulses without causing unconsciousness. They act by reversibly binding to fast sodium channels from within nerve fibers, thereby preventing sodium from entering the fibres, stabilising the cell membrane and preventing action potential propagation. Each of the local anesthetics have the suffix "-caine" in their names.

Local anesthetics can be either ester- or amide-based. Ester local anesthetics (such as procaine, amethocaine, cocaine, benzocaine, tetracaine) are generally unstable in solution and fast-acting, are rapidly metabolised by cholinesterases in the blood plasma and liver, and more commonly induce allergic reactions.

Amide local anesthetics (such lidocaine, prilocaine, bupivicaine, levobupivacaine, ropivacaine, mepivacaine, dibucaine and etidocaine) are generally heat-stable, with a long shelf life (around two years). Amides have a slower onset and longer half-life than ester anesthetics, and are usually racemic mixtures, with the exception of levobupivacaine (which is S(-) -bupivacaine) and ropivacaine (S(-)-ropivacaine). Amides are generally used within regional and epidural or spinal techniques, due to their longer duration of action, which provides adequate analgesia for surgery, labor, and symptomatic relief.

Only preservative-free local anesthetic agents may be injected intrathecally.

Pethidine also has local anesthetic properties, in addition to its opioid effects.

General anesthetics

Inhaled agents

- Desflurane
- Enflurane
- Halothane
- Isoflurane
- Methoxyflurane
- Nitrous oxide
- Sevoflurane
- Xenon (rarely used)

Volatile agents are specially formulated organic liquids that evaporate readily into vapors, and are given by inhalation for induction or maintenance of general anesthesia. Nitrous oxide and xenon are gases at room temperature rather than liquids, so they are not considered volatile agents. The ideal anesthetic vapor or gas should be non-flammable, non-explosive, and lipid-soluble. It should possess low blood gas solubility, have no end-organ (heart, liver, kidney) toxicity or side-effects, should not be metabolized, and should not be an irritant to the respiratory pathways of the patient.

No anaesthetic agent currently in use meets all these requirements, nor can any anaesthetic agent be considered safe. There are inherent risks and drug interactions that are specific to each and every patient. The agents in widespread current use are isoflurane, desflurane, sevoflurane, and nitrous oxide. Nitrous oxide is a common adjuvant gas, making it one of the most long-lived drugs still in current use. Because of its low potency, it cannot produce anesthesia on its own but is frequently combined with other agents. Halothane, an agent introduced in the 1950s, has been almost completely replaced in modern anesthesia practice by newer agents because of its shortcomings. Partly because of its side effects, enflurane never gained widespread popularity.

In theory, any inhaled anesthetic agent can be used for induction of general anesthesia. However, most of the halogenated anesthetics are irritating to the airway, perhaps leading to coughing, laryngospasm and overall difficult inductions. For this reason, the most frequently used agent for inhalational induction is sevoflurane. All of the volatile agents can be used alone or in combination with other medications to maintain anesthesia (nitrous oxide is not potent enough to be used as a sole agent). Volatile agents are frequently compared in terms of potency, which is inversely proportional to the minimum alveolar concentration. Potency is directly related to lipid solubility.

This is known as the Meyer-Overton hypothesis. However, certain pharmacokinetic properties of volatile agents have become another point of comparison. Most important of those properties is known as the blood/gas partition coefficient. This concept refers to the relative solubility of a given agent in blood. Those agents with a lower blood solubility (i.e., a lower blood–gas partition coefficient; e.g., desflurane) give the anesthesia provider greater rapidity in titrating the depth of anesthesia, and permit a more rapid emergence from the anesthetic state upon discontinuing their administration. In fact, newer volatile agents (e.g., sevoflurane, desflurane) have been popular not due to their potency (minimum alveolar concentration), but due to their versatility for a faster emergence from anesthesia, thanks to their lower blood–gas partition coefficient.

Intravenous agents (non-opioid)

While there are many drugs that can be used intravenously to produce anesthesia or sedation, the most common are:

- Barbiturates
- Amobarbital (trade name: Amytal)
- Methohexital (trade name: Brevital)
- Thiamylal (trade name: Surital)
- Thiopental (trade name: Penthothal, referred to as thiopentone in the UK)
- Benzodiazepines
- Diazepam

- Lorazepam
- Midazolam
- Etomidate
- Ketamine
- Propofol

The two barbiturates mentioned above, thiopental and methohexital, are ultra-short-acting, and are used to induce and maintain anesthesia. However, though they produce unconsciousness, they provide no analgesia (pain relief) and must be used with other agents. Benzodiazepines can be used for sedation before or after surgery and can be used to induce and maintain general anesthesia. When benzodiazepines are used to induce general anesthesia, midazolam is preferred. Benzodiazepines are also used for sedation during procedures that do not require general anesthesia. Like barbiturates, benzodiazepines have no pain-relieving properties. Propofol is one of the most commonly used intravenous drugs employed to induce and maintain general anesthesia. It can also be used for sedation during procedures or in the ICU. Like the other agents mentioned above, it renders patients unconscious without producing pain relief. Because of its favorable physiological effects, "etomidate has been primarily used in sick patients". Ketamine is infrequently used in anesthesia because of the unpleasant experiences that sometimes occur on emergence from anesthesia, which include "vivid dreaming, extracorporeal experiences, and illusions." However, like etomidate it is frequently used in emergency settings and with sick patients because it produces fewer adverse physiological effects.Unlike the intravenous anesthetic drugs previously mentioned, ketamine produces profound pain relief, even in doses lower than those that induce general anesthesia. Also unlike the other anesthetic agents in this section, patients who receive ketamine alone appear to be in a cataleptic state, unlike other states of anesthesia that resemble normal sleep. Ketamine-anesthetized patients have profound analgesia but keep their eyes open and maintain many reflexes.

NEONATAL AND INFANT NEUROTOXICITY CONCERNS

Concerns have been raised as to the safety of general anesthetics, in particular ketamine and isoflurane in neonates and young children due to significant neurodegeneration. The risk of neurodegeneration is increased in combination of these agents with nitrous oxide and benzodiazepines such as midazolam. This has led to the FDA and other bodies to take steps to investigate these concerns. These concerns have arisen from animal studies involving rats and non-human primates.

Research has found that anesthetics which enhance GABA or block NMDA can precipitate neuronal cell death in these animals. The developing central nervous system is most vulnerable to these potential neurotoxic effects during the last trimester of pregnancy and shortly after birth. Melatonin, a free oxygen radical scavenger and indirect antioxidant is known to reduce the toxicity of a range of drugs has been found in a rat study to reduce the neurotoxicity of anesthetic agents to the early developing brain. Recent research in animals has found that all sedatives and anesthetics cause extensive neurodegeneration in the developing brain. There is also some evidence in humans that surgery and exposure to anesthetics in the early developmental stages causes persisting learning deficits.

Intravenous opioid analgesic agents

While opioids can produce unconsciousness, they do so unreliably and with significant side effects. So, while they are rarely used to induce anesthesia, they are frequently used along with other agents such as intravenous non-opioid anesthetics or inhalational anesthetics. Furthermore, they are used to relieve pain of patients before, during, or after surgery. The following opioids have short onset and duration of action and are frequently used during general anesthesia:

- Alfentanil
- Fentanyl
- Remifentanil
- Sufentanil, which is not available in Australia.

The following agents have longer onset and duration of action and are frequently used for post-operative pain relief:

- Buprenorphine
- Butorphanol
- Diamorphine, also known as heroin, not available for use as an analgesic in any country but the UK.
- Hydromorphone
- Levorphanol
- Pethidine, also called meperidine in North America.
- Methadone
- Morphine
- Nalbuphine

- Oxycodone, not available intravenously in U.S.
- Oxymorphone
- Pentazocine

Muscle relaxants

Muscle relaxants do not render patients unconscious or relieve pain. Instead, they are sometimes used after a patient is rendered unconscious (induction of anesthesia) to facilitate intubation or surgery by paralyzing skeletal muscle.

- Depolarizing muscle relaxants
- Succinylcholine (also known as suxamethonium in the UK, New Zealand, Australia and other countries, "Celokurin" or "celo" for short in Europe)
- Decamethonium
- Non-depolarizing muscle relaxants
- Short acting
- Mivacurium
- Rapacuronium
- Intermediate acting
- Atracurium
- Cisatracurium
- Rocuronium
- Vecuronium
- Long acting
- Alcuronium
- Doxacurium
- Gallamine
- Metocurine
- Pancuronium
- Pipecuronium
- Tubocurarine

Adverse effects

- Depolarizing Muscle Relaxants i.e. Suxamethonium

- Hyperkalemia – A small rise of 0.5 mmol/l occurs normally, this is of little consequence unless potassium is already raised such as in renal failure
- Hyperkalemia – Exaggerated potassium release in burn patients (occurs from 24 hours after injury, lasting for up to 2 years), neuromuscular disease and paralyzed (quadraplegic, paraplegic) patients. The mechanism is reported to be through upregulation of acetylcholine receptors in those patient populations with increased efflux of potassium from inside muscle cells. May cause life-threatening arrhythmia
- Muscle aches, commoner in young muscular patients who mobilize soon after surgery
- Bradycardia, especially if repeat doses are given
- Malignant hyperthermia, a potentially life-threatening condition in susceptible patients
- Suxamethonium Apnea, a rare genetic condition leading to prolonged duration of neuromuscular blockade, this can range from 20 minutes to a number of hours. Not dangerous as long as it is recognized and the patient remains intubated and sedated, there is the potential for awareness if this does not occur.
- Anaphylaxis
- Non-depolarizing Muscle Relaxants
- Histamine release e.g. Atracurium & Mivacurium
- Anaphylaxis

Another potentially disturbing complication where neuromuscular blockade is employed is 'anesthesia awareness'. In this situation, patients paralyzed may awaken during their anesthesia, due to an inappropriate decrease in the level of drugs providing sedation or pain relief. If this fact is missed by the anesthesia provider, the patient may be aware of their surroundings, but be incapable of moving or communicating that fact. Neurological monitors are increasingly available that may help decrease the incidence of awareness. Most of these monitors use proprietary algorithms monitoring brain activity via evoked potentials. Despite the widespread marketing of these devices many case reports exist in which awareness under anesthesia has occurred despite apparently adequate anesthesia as measured by the neurologic monitor.

Intravenous reversal agents

- Flumazenil, reverses the effects of benzodiazepines
- Naloxone, reverses the effects of opioids
- Neostigmine, helps reverse the effects of non-depolarizing muscle relaxants
- Sugammadex, new agent that is designed to bind Rocuronium therefore terminating its action

Anesthesia

Anesthesia or anaesthesia (from Greek "without sensation") is a state of controlled, temporary loss of sensation or awareness that is induced for medical purposes. It may include analgesia (relief from or prevention of pain), paralysis (muscle relaxation), amnesia (loss of memory), or unconsciousness. A patient under the effects of anesthetic drugs is referred to as being anesthetized.

Anesthesia enables the painless performance of medical procedures that would otherwise cause severe or intolerable pain to an unanesthetized patient, or would otherwise be technically unfeasible. Three broad categories of anesthesia exist:

- General anesthesia suppresses central nervous system activity and results in unconsciousness and total lack of sensation.
- Sedation suppresses the central nervous system to a lesser degree, inhibiting both anxiety and creation of long-term memories without resulting in unconsciousness.
- Regional and local anesthesia, which block transmission of nerve impulses from a specific part of the body. Depending on the situation, this may be used either on its own (in which case the patient remains conscious), or in combination with general anesthesia or sedation. Drugs can be targeted at peripheral nerves to anesthetize an isolated part of the body only, such as numbing a tooth for dental work or using a nerve block to inhibit sensation in an entire limb. Alternatively, epidural or spinal anesthesia can be performed in the region of the central nervous system itself, suppressing all incoming sensation from nerves outside the area of the block.

In preparing for a medical procedure, the clinician chooses one or more drugs to achieve the types and degree of anesthesia characteristics appropriate for the type of procedure and the particular patient. The types of drugs used include general anesthetics, local anesthetics, hypnotics, sedatives, neuromuscular-blocking drugs, narcotics, and analgesics.

Risks during and following anesthesia are difficult to quantify, since many may be related to a variety of factors related to anesthesia itself, the nature of the procedure being performed and the patient's medical health. Examples of major risks include death, heart attack and pulmonary embolism whereas minor risks can include postoperative nausea and vomiting and hospital readmission. Of these factors, the person's health prior to the procedure(stratified by the ASA physical status classification system) has the greatest bearing on the probability of a complication occurring. Patients typically wake within minutes of anesthesia being terminated and regain their senses within hours.

Medical uses

The purpose of anesthesia can be distilled down to three basic goals or endpoints:

- Hypnosis (a temporary loss of consciousness and with it a loss of memory. In a pharmacological context, the word hypnosis usually has this technical meaning, in contrast to its more familiar lay or psychological meaning of an altered state of consciousness not necessarily caused by drugs—see hypnosis).
- Analgesia (lack of sensation which also blunts autonomic reflexes)
- Muscle relaxation

Different types of anesthesia affect the endpoints differently. Regional anesthesia, for instance, affects analgesia; benzodiazepine-type sedatives (used for sedation, or "twilight anesthesia") favor amnesia; and general anesthetics can affect all of the endpoints. The goal of anesthesia is to achieve the endpoints required for the given surgical procedure with the least risk to the patient.

To achieve the goals of anesthesia, drugs act on different but interconnected parts of the nervous system. Hypnosis, for instance, is generated through actions on the nuclei in the brain and is similar to the activation of sleep. The effect is to make people less aware and less reactive to noxious stimuli.

Loss of memory (amnesia) is created by action of drugs on multiple (but specific) regions of the brain. Memories are created as either declarative or non-declarative memories in several stages (short-term, long-term, long-lasting) the strength of which is determined by the strength of connections between neurons termed synaptic plasticity. Each anesthetic produces amnesia through unique effects on memory formation at variable doses. Inhalational anesthetics will reliably produce amnesia through general suppression of the nuclei at doses below those required for loss of consciousness. Drugs like midazolam produce amnesia through different pathways by blocking the formation of long-term memories. Tied closely to the concepts of amnesia and hypnosis is the concept of consciousness. Consciousness is the higher order process that synthesizes information. For instance, the "sun" conjures up feelings, memories and a sensation of warmth rather than a description of a round, orange warm ball seen in the sky for part of a 24 hour cycle. Likewise, a person can have dreams (a state of subjective consciousness) during anesthetic or have consciousness of the procedure despite having no indication of it under anesthetic. It is estimated that 22% of people dream during general anesthesia and 1 or 2 cases per 1000 have some consciousness termed "awareness during general anesthesia".

Techniques

Safe anesthesia care depends greatly on well-functioning teams of highly trained healthcare workers. The medical specialty centred around anesthesia is called anesthesiology, and medical doctors who practise it are termed anesthesiologists. Non-medical healthcare workers involved in anesthesia provision have varying titles and roles depending on the jurisdiction, and include nurse anesthetists, anesthetic nurses, anesthesiologist assistants, anaesthetic technicians, physicians' assistants (anaesthesia), operating department practitioners and anesthesia technologists. International standards for the safe practice of anesthesia, jointly endorsed by the World Health Organization and the World Federation of Societies of Anaesthesiologists, highly recommend that anesthesia should be provided, overseen or led by anesthesiologists, with the exception of minimal sedation or superficial procedures performed under local anesthesia. A trained, vigilant anesthesia provider should continually care for the patient; where the provider is not an anesthesiologist, they should be locally directed and supervised by an anesthesiologist, and in countries or settings where this is not feasible, care should be led by the most qualified local individual within a regional or national anesthesiologist-led framework. The same minimum standards for patient safety apply regardless of the provider, including continuous clinical and biometric monitoring of tissue oxygenation, perfusion and blood pressure; confirmation of correct placement of airway management devices by auscultation and carbon dioxide detection; use of the WHO Surgical Safety Checklist; and safe onward transfer of the patient's care following the procedure.

Anesthesia is unique in that it is not a direct means of treatment; rather, it allows others to do things that may treat, diagnose, or cure an ailment which would otherwise be painful or complicated. The best anesthetic, therefore, is the one with the lowest risk to the patient that still achieves the endpoints required to complete the procedure. The first stage in anesthesia is the pre-operative risk assessment consisting of the medical history, physical examination and lab tests. Diagnosing a person's pre-operative physical status allows the clinician to minimize anesthetic risks. A well completed medical history will arrive at the correct diagnosis 56% of the time which increases to 73% with a physical examination. Lab tests help in diagnosis but only in 3% of cases, underscoring the need for a full history and physical examination prior to anesthetics. Incorrect pre-operative assessments or preparations are the root cause of 11% of all adverse anesthetic events.

One part of the risk assessment is based on the patients' health. The American Society of Anesthesiologists has developed a six-tier scale that stratifies the patient's pre-operative physical state. It is called the ASA physical status. The scale assesses risk as the patient's general health relates to an anesthetic.

The more detailed pre-operative medical history aims to discover genetic disorders (such as malignant hyperthermia or pseudocholinesterase deficiency), habits (tobacco, drug and alcohol use), physical attributes (such as obesity or a difficult airway) and any coexisting diseases (especially cardiac and respiratory diseases) that might impact the anesthetic. The physical examination helps quantify the impact of anything found in the medical history in addition to lab tests.

Aside from the generalities of the patients health assessment, an evaluation of specific factors as they relate to the surgery also need to be considered for anesthesia. For instance, anesthesia during childbirth must consider not only the mother but the baby. Cancers and tumors that occupy the lungs or throat create special challenges to general anesthesia. After determining the health of the person undergoing anesthetic and the endpoints that are required to complete the procedure, the type of anesthetic can be selected. Choice of surgical method and anesthetic technique aims to reduce risk of complications, shorten time needed for recovery and minimise the surgical stress response.

General anesthesia

Anesthesia is a combination of the endpoints (discussed above) that are reached by drugs acting on different but overlapping sites in the central nervous system. General anesthesia (as opposed to sedation or regional anesthesia) has three main goals: lack of movement (paralysis), unconsciousness, and blunting of the stress response. In the early days of anesthesia, anesthetics could reliably achieve the first two, allowing surgeons to perform necessary procedures, but many patients died because the extremes of blood pressure and pulse caused by the surgical insult were ultimately harmful. Eventually, the need for blunting of the surgical stress response was identified by Harvey Cushing, who injected local anesthetic prior to hernia repairs. This led to the development of other drugs that could blunt the response leading to lower surgical mortality rates.

A vaporizer holds a liquid anesthetic and converts it to gas for inhalation (in this case sevoflurane)

The most common approach to reach the endpoints of general anesthesia is through the use of inhaled general anesthetics. Each anesthetic has its own potency which is correlated to its solubility in oil. This relationship exists because the drugs bind directly to cavities in proteins of the central nervous system, although several theories of general anesthetic action have been described. Inhalational anesthetics are thought to exact their effects on different parts of the central nervous system. For instance, the immobilizing effect of inhaled anesthetics results from an effect on the spinal cord whereas sedation, hypnosis and amnesia involve sites in the brain. The potency of an inhalational anesthetic is quantified by its minimum alveolar concentration or MAC. The MAC is the percentage dose of anesthetic that will prevent a response to painful stimulus in 50% of subjects. The higher the MAC, generally, the less potent the anesthetic.

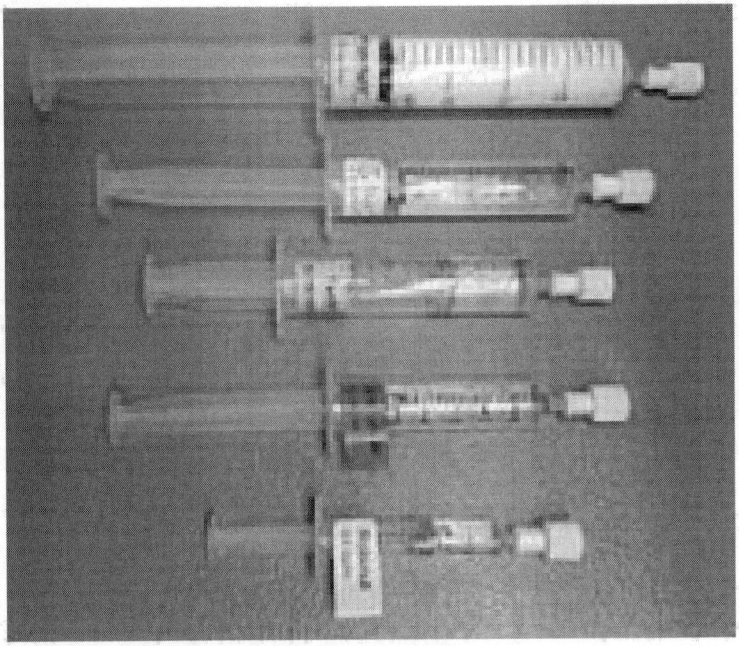

Syringes prepared with medications that are expected to be used during an operation under general anesthesia maintained by sevoflurane gas:

- Propofol, an hypnotic
- Ephedrine, in case of hypotension
- Fentanyl, for analgesia
- Atracurium, for neuromuscular blockade
- Glycopyrronium bromide (here under trade name "Robinul"), reducing secretions

The ideal anesthetic drug would provide hypnosis, amnesia, analgesia, and muscle relaxation without undesirable changes in blood pressure, pulse or breathing. In the 1930s, physicians started to augment inhaled general anesthetics with intravenous general anesthetics. The drugs used in combination offered a better risk profile to the person under anesthesia and a quicker recovery. A combination of drugs was later shown to result in lower odds of dying in the first 7 days after anesthetic. For instance, propofol (injection) might be used to start the anesthetic, fentanyl (injection) used to blunt the stress response, midazolam (injection) given to ensure amnesia and sevoflurane (inhaled) during the procedure to maintain the effects. More recently, several intravenous drugs have been developed which, if desired, allow inhaled general anesthetics to be avoided completely.

Equipment

The core instrument in an inhalational anesthetic delivery system is an anesthetic machine. It has vaporizers, ventilators, an anesthetic breathing circuit, waste gas scavenging system and pressure gauges. The purpose of the anesthetic machine is to provide anesthetic gas at a constant pressure, oxygen for breathing and to remove carbon dioxide or other waste anesthetic gases. Since inhalational anesthetics are flammable, various checklists have been developed to confirm that the machine is ready for use, that the safety features are active and the electrical hazards are removed. Intravenous anesthetic is delivered either by bolus doses or an infusion pump. There are also many smaller instruments used in airway management and monitoring the patient. The common thread to modern machinery in this field is the use of fail-safe systems that decrease the odds of catastrophic misuse of the machine.

Monitoring

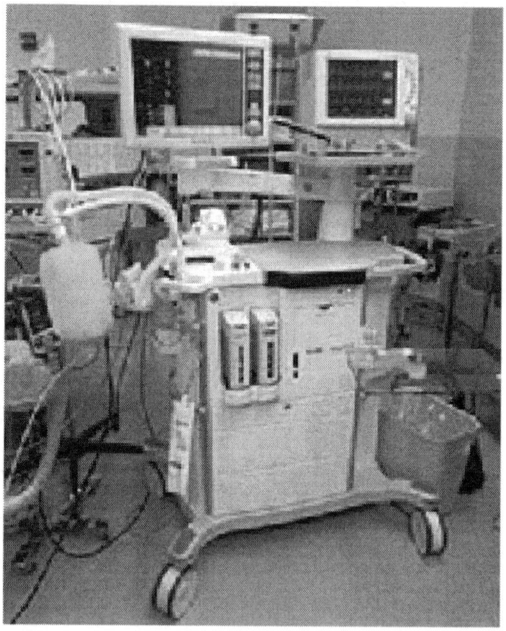

fAn anesthetic machine with integrated systems for monitoring of several vital parameters.

Patients under general anesthesia must undergo continuous physiological monitoring to ensure safety. In the US, the American Society of Anesthesiologists (ASA) has established minimum monitoring guidelines for patients receiving general anesthesia, regional anesthesia, or sedation. These include electrocardiography

(ECG), heart rate, blood pressure, inspired and expired gases, oxygen saturation of the blood (pulse oximetry), and temperature. In the UK the Association of Anaesthetists (AAGBI) have set minimum monitoring guidelines for general and regional anesthesia. For minor surgery, this generally includes monitoring of heart rate, oxygen saturation, blood pressure, and inspired and expired concentrations for oxygen, carbon dioxide, and inhalational anesthetic agents. For more invasive surgery, monitoring may also include temperature, urine output, blood pressure, central venous pressure, pulmonary artery pressure and pulmonary artery occlusion pressure, cardiac output, cerebral activity, and neuromuscular function. In addition, the operating room environment must be monitored for ambient temperature and humidity, as well as for accumulation of exhaled inhalational anesthetic agents, which might be deleterious to the health of operating room personnel.

Sedation

Further information: Sedation

Sedation (also referred to as dissociative anesthesia or twilight anesthesia) creates hypnotic, sedative, anxiolytic, amnesic, anticonvulsant, and centrally produced muscle-relaxing properties. From the perspective of the person giving the sedation, the patient appears sleepy, relaxed and forgetful, allowing unpleasant procedures to be more easily completed. Sedatives such as benzodiazepines are usually given with pain relievers (such as narcotics, or local anesthetics or both) because they do not, by themselves, provide significant pain relief.

From the perspective of the person receiving a sedative, the effect is a feeling of general relaxation, amnesia (loss of memory) and time passing quickly. Many drugs can produce a sedative effect including benzodiazepines, propofol, thiopental, ketamine and inhaled general anesthetics. The advantage of sedation over a general anesthetic is that it generally does not require support of the airway or breathing (no tracheal intubation or mechanical ventilation) and can have less of an effect on the cardiovascular system which may add to a greater margin of safety in some patients.

Regional anesthesia

Backflow of cerebrospinal fluidthrough a spinal needle after puncture of the arachnoid materduring spinal anesthesia

When pain is blocked from a part of the body using local anesthetics, it is generally referred to as regional anesthesia. There are many types of regional anesthesia either by injecting into the tissue itself, a vein that feeds the area or around a nerve trunk that supplies sensation to the area. The latter are called nerve blocks and are divided into peripheral or central nerve blocks.

The following are the types of regional anesthesia:

- Infiltrative anesthesia: a small amount of local anesthetic is injected in a small area to stop any sensation (such as during the closure of a laceration, as a continuous infusion or "freezing" a tooth). The effect is almost immediate.
- Peripheral nerve block: local anesthetic is injected near a nerve that provides sensation to particular portion of the body. There is significant variation in the speed of onset and duration of anesthesia depending on the potency of the drug (e.g. Mandibular block).
- Intravenous regional anesthesia (also called a Bier block): dilute local anesthetic is infused to a limb through a vein with a tourniquet placed to prevent the drug from diffusing out of the limb.
- Central nerve blockade: Local anesthetic is injected or infused in or around a portion of the central nervous system (discussed in more detail below in Spinal, epidural and caudal anesthesia).
- Topical anesthesia: local anesthetics that are specially formulated to diffuse through the mucous membranes or skin to give a thin layer of analgesia to an area (e.g. EMLA patches).
- Tumescent anesthesia: a large amount of very dilute local anesthetics are injected into the subcutaneous tissues during liposuction.
- Systemic local anesthetics: local anesthetics are given systemically (orally or intravenous) to relieve neuropathic pain

Nerve blocks

When local anesthetic is injected around a larger diameter nerve that transmits sensation from an entire region it is referred to as a nerve block or regional nerve blockade. Nerve blocks are commonly used in dentistry, when the mandibular nerve is blocked for procedures on the lower teeth. With larger diameter nerves (such as the interscalene block for upper limbs or psoas compartment block for lower limbs) the nerve and position of the needle is localized with ultrasound or electrical stimulation. The use of ultrasound may reduce complication rates and improve quality, performance time, and time to onset of blocks. Because of the large amount of local anesthetic required to affect the nerve, the maximum dose of local anesethetic has to be considered. Nerve blocks are also used as a continuous infusion, following major surgery such as knee, hip and shoulder replacement surgery, and may be associated with lower complications. Nerve blocks are also associated with a lower risk of neurologic complications compared to the more central epidural or spinal neuraxial blocks.

Spinal, epidural and caudal anesthesia

Central neuraxial anesthesia is the injection of local anesthetic around the spinal cord to provide analgesia in the abdomen, pelvis or lower extremities. It is divided into either spinal (injection into the subarachnoid space), epidural (injection outside of the subarachnoid space into the epidural space) and caudal (injection into the cauda equina or tail end of the spinal cord). Spinal and epidural are the most commonly used forms of central neuraxial blockade.

Spinal anesthesia is a "one-shot" injection that provides rapid onset and profound sensory anesthesia with lower doses of anesethetic, and is usually associated with neuromuscular blockade (loss of muscle control). Epidural anesthesia uses larger doses of anesthetic infused through an indwelling catheter which allows the anesthetic to be augmented should the effects begin to dissipate. Epidural anesthesia does not typically affect muscle control. Because central neuraxial blockade causes arterial and vasodilation, a drop in blood pressure is common. This drop is largely dictated by the venous side of the circulatory systemwhich holds 75% of the circulating blood volume. The physiologic effects are much greater when the block is placed above the 5th thoracic vertebra. An ineffective block is most often due to inadequate anxiolysis or sedation rather than a failure of the block itself.

Acute pain management

Pain that is well managed during and immediately after surgery improves the health of patients (by decreasing physiologic stress) and the potential for chronic pain. Nociception (pain sensation) is not hard-wired into the body. Instead, it is a dynamic process wherein persistent painful stimuli can sensitize the system and either make pain management difficult or promote the development of chronic pain. For this reason, preemptive acute pain management may reduce both acute and chronic pain and is tailored to the surgery, the environment in which it is given (in-patient/out-patient) and the individual patient.

Pain management is classified into either pre-emptive or on-demand. On-demand pain medications typically include either opioid or non-steroidal anti-inflammatory drugs but can also make use of novel approaches such as inhaled nitrous oxide or ketamine. On demand drugs can be administered by a clinician ("as needed drug orders") or by the patient using patient-controlled analgesia (PCA). PCA has been shown to provide slightly better pain control and increased patient satisfaction when compared with conventional methods. Common preemptive approaches include epidural neuraxial blockade or nerve blocks. One review which looked at pain control after abdominal aortic surgery found that epidural blockade provides better pain relief (especially during movement) in the period up to three postoperative days. It reduces the duration of postoperative tracheal intubation by roughly half. The occurrence of prolonged postoperative mechanical ventilation and myocardial infarction is also reduced by epidural analgesia.

Risks and complications

Risks and complications as they relate to anesthesia are classified as either morbidity (a disease or disorder that results from anesthesia) or mortality (death that results from anesthesia). Quantifying how anesthesia contributes to morbidity and mortality can be difficult because a person's health prior to surgery and the complexity of the surgical procedure can also contribute to the risks.

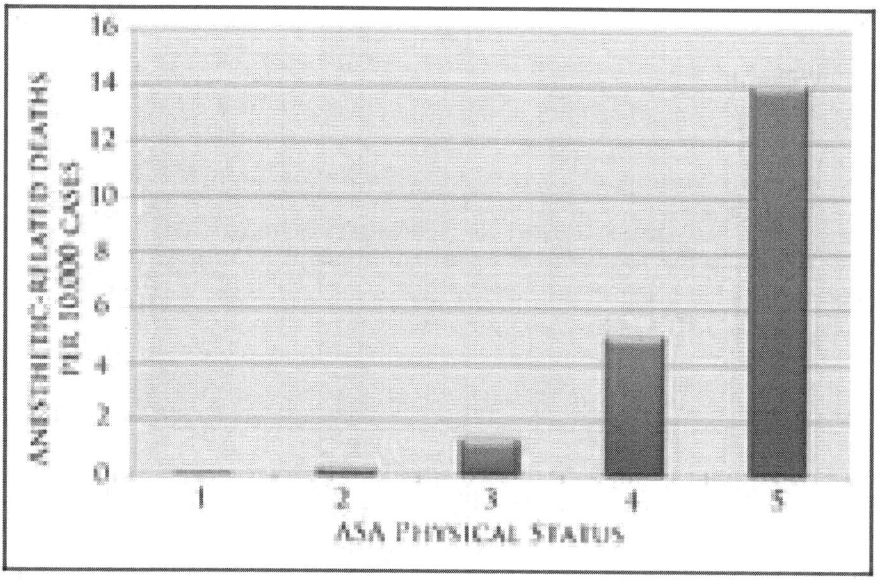

Anesthesia-related deaths by ASA status

Prior to the introduction of anesthesia in the early 19th century, the physiologic stress from surgery caused significant complications and many deaths from shock. The faster the surgery was, the lower the rate of complications (leading to reports of very quick amputations). The advent of anesthesia allowed more complicated and life-saving surgery to be completed, decreased the physiologic stress of the surgery, but added an element of risk. It was two years after the introduction of ether anesthetics that the first death directly related to the use of anesthesia was reported.

Morbidity can be major (myocardial infarction, pneumonia, pulmonary embolism, renal failure/insufficiency, postoperative cognitive dysfunction and allergy) or minor (minor nausea, vomiting, readmission). There is usually overlap in the contributing factors that lead to morbidity and mortality between the health of the patient, the type of surgery being performed and the anesthetic. To understand the relative risk of each contributing factor, consider that the rate of deaths totally attributed to the patient's health is 1:870. Compare that to the rate of deaths totally attributed to surgical factors (1:2860) or anesthesia alone (1:185,056) illustrating that the single greatest factor in anesthetic mortality is the health of the patient. These statistics can also be compared to the first such study on mortality in anesthesia from 1954, which reported a rate of death from all causes at 1:75 and a rate attributed to anesthesia alone at 1:2680. Direct comparisons between mortality statistics cannot reliably be made over time and across countries because of differences in the stratification of risk factors, however, there is evidence that anesthetics have made a significant improvement in safety but to what degree is uncertain.

Rather than stating a flat rate of morbidity or mortality, many factors are reported as contributing to the relative risk of the procedure and anesthetic combined. For instance, an operation on a person who is between the ages of 60–79 years old places the patient at 2.3 times greater risk than someone less than 60 years old. Having an ASA score of 3, 4 or 5 places the person at 10.7 times greater risk than someone with an ASA score of 1 or 2. Other variables include age greater than 80 (3.3 times risk compared to those under 60), gender (females have a lower risk of 0.8), urgency of the procedure (emergencies have a 4.4 times greater risk), experience of the person completing the procedure (less than 8 years experience and/or less than 600 cases have a 1.1 times greater risk) and the type of anesthetic (regional anesthetics are lower risk than general anesthetics). Obstetrical, the very young and the very old are all at greater risk of complication so extra precautions may need to be taken.

On 14 December 2016 the Food and Drug Administration issued a Public Safety Communication warning that "repeated or lengthy use of general anesthetic and sedation drugs during surgeries or procedures in children younger than 3 years or in pregnant women during their third trimester may affect the development of children's brains." The warning was criticized by the American College of Obstetricians and Gynecologists, which pointed out the absence of direct evidence regarding use in pregnant women and the possibility that "this warning could inappropriately dissuade providers from providing medically indicated care during pregnancy." Patient advocates noted that a randomized clinical trial would be unethical, that the mechanism of injury is well-established in animals, and that studies had shown exposure to multiple uses of anesthetic significantly increased the risk of developing learning disabilities in young children, with a hazard ratio of 2.12 (95% confidence interval, 1.26–3.54).

RECOVERY

The immediate time after anesthesia is called emergence. Emergence from general anesthesia or sedation requires careful monitoring because there is still a risk of complication. Nausea and vomiting are reported at 9.8% but will vary with the type of anesthetic and procedure. There is a need for airway support in 6.8%, there can be urinary retention (more common in those over 50 years of age) and hypotension in 2.7%. Hypothermia, shivering and confusion are also common in the immediate post-operative period because of the lack of muscle movement (and subsequent lack of heat production) during the procedure.

Postoperative cognitive dysfunction (also known as POCD and post-anesthetic confusion) is a disturbance in cognition after surgery. It may also be variably used to describe emergence delirium (immediate post-operative confusion) and early cognitive dysfunction (diminished cognitive function in the first post-operative week). Although the three entities (delirium, early POCD and long-term POCD) are separate, the presence of delirium post-operatively predicts the presence of early POCD. There does not appear to be an association between delirium or early POCD and long-term POCD. According to a recent study conducted at the David Geffen School of Medicine at UCLA, the brain navigates its way through a series of activity clusters, or "hubs" on its way back to consciousness. Dr. Andrew Hudson, an assistant professor in anesthesiology states, "Recovery from anesthesia is not simply the result of the anesthetic 'wearing off,' but also of the brain finding its way back through a maze of possible activity states to those that allow conscious experience. Put simply, the brain reboots itself."

Long-term POCD is a subtle deterioration in cognitive function, that can last for weeks, months, or longer. Most commonly, relatives of the person report a lack of attention, memory and loss of interest in activities previously dear to the person (such as crosswords). In a similar way, people in the workforce may report an inability to complete tasks at the same speed they could previously. There is good evidence that POCD occurs after cardiac surgery and the major reason for its occurrence is the formation of microemboli. POCD also appears to occur in non-cardiac surgery. Its causes in non-cardiac surgery are less clear but older age is a risk factor for its occurrence.

History

The first attempts at general anesthesia were probably herbal remedies administered in prehistory. Alcohol is one of the oldest known sedatives and it was used in ancient Mesopotamia thousands of years ago. The Sumerians are said to have cultivated and harvested the opium poppy (*Papaver somniferum*) in lower Mesopotamia as early as 3400 BC.

The ancient Egyptians had some surgical instruments, as well as crude analgesics and sedatives, including possibly an extract prepared from the mandrake fruit. Bian Que(Chinese: AbJž, Wade–Giles: Pien Ch'iao, c. 300 BC) was a legendary Chinese internist and surgeon who reportedly used general anesthesia for surgical procedures.

Throughout Europe, Asia, and the Americas a variety of Solanum species containing potent tropane alkaloids were used for anesthesia. In 13th-century Italy, Theodoric Borgognoniused similar mixtures along with opiates to induce unconsciousness, and treatment with the combined alkaloids proved a mainstay of anesthesia until the nineteenth century. Local anesthetics were used in Inca civilization where shamans chewed coca leaves and performed operations on the skull while spitting into the wounds they had inflicted to anesthetize. Cocaine was later isolated and became the first effective local anesthetic. It was first used in 1859 by Karl Koller, at the suggestion of Sigmund Freud, in eye surgeryin 1884. German surgeon August Bier (1861–1949) was the first to use cocaine for intrathecal anesthesia in 1898. Romanian surgeon Nicolae Racoviceanu-Piteºti (1860–1942) was the first to use opioids for intrathecal analgesia; he presented his experience in Paris in 1901.

Early Arab writings mention anesthesia by inhalation. This idea was the basis of the "soporific sponge" ("sleep sponge"), introduced by the Salerno school of medicine in the late twelfth century and by Ugo Borgognoni (1180–1258) in the thirteenth century. The sponge was promoted and described by Ugo's son and fellow surgeon, Theodoric Borgognoni(1205–1298). In this anesthetic method, a sponge was soaked in a dissolved solution of opium, mandragora, hemlock juice, and other substances. The sponge was then dried and stored; just before surgery the sponge was moistened and then held under the patient's nose. When all went well, the fumes rendered the patient unconscious.

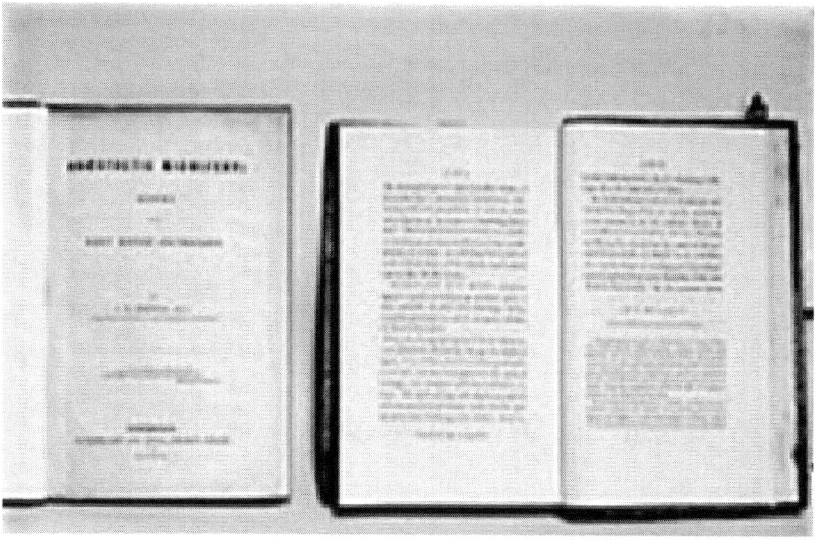

Sir Humphry Davy's Researches chemical and philosophical: chiefly concerning nitrous oxide (1800), pages 556 and 557 (right), outlining potential anesthetic properties of nitrous oxide in relieving pain during surgery

The most famous anesthetic, ether, may have been synthesized as early as the 8th century, but it took many centuries for its anesthetic importance to be appreciated, even though the 16th century physician and polymath Paracelsus noted that chickens made to breathe it not only fell asleep but also felt no pain. By the early 19th century, ether was being used by humans, but only as a recreational drug.

Meanwhile, in 1772, English scientist Joseph Priestley discovered the gas nitrous oxide. Initially, people thought this gas to be lethal, even in small doses, like some other nitrogen oxides. However, in 1799, British chemist and inventor Humphry Davy decided to find out by experimenting on himself. To his astonishment he found that nitrous oxide made him laugh, so he nicknamed it "laughing gas". In 1800 Davy wrote about the potential anesthetic properties of nitrous oxide in relieving pain during surgery, but nobody at that time pursued the matter any further.

American physician Crawford W. Long noticed that his friends felt no pain when they injured themselves while staggering around under the influence of diethyl ether. He immediately thought of its potential in surgery. Conveniently, a participant in one of those "ether frolics", a student named James Venable, had two small tumors he wanted excised. But fearing the pain of surgery, Venable kept putting the operation off. Hence, Long suggested that he have his operation while under the influence of ether. Venable agreed, and on 30 March 1842 he underwent a painless operation. However, Long did not announce his discovery until 1849.

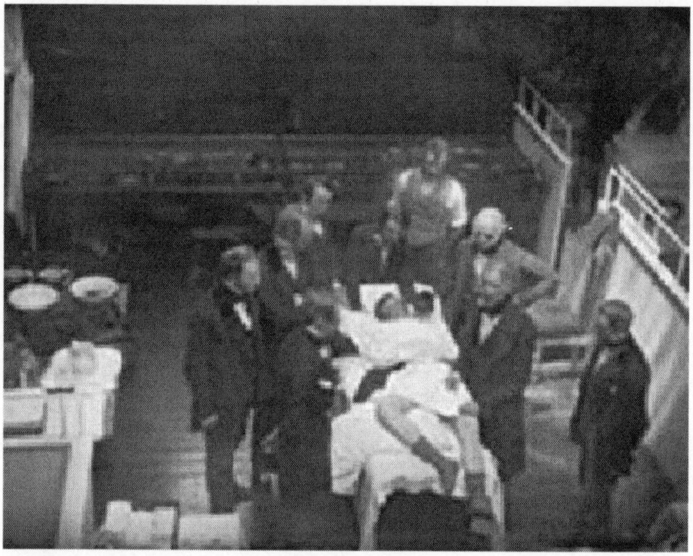

Contemporary re-enactment of Morton's 16 October 1846, etheroperation; daguerrotype by Southworth & Hawes

MORTON'S ETHER INHALER

Horace Wells conducted the first public demonstration of the inhalational anesthetic at the Massachusetts General Hospital in Boston in 1845. However, the nitrous oxide was improperly administered and the patient cried out in pain. On 16 October 1846, Boston dentist William Thomas Green Morton gave a successful demonstration using diethyl ether to medical students at the same venue. Morton, who was unaware of Long's previous work, was invited to the Massachusetts General Hospital to demonstrate his new technique for painless surgery. After Morton had induced anesthesia, surgeon John Collins Warren removed a tumor from the neck of Edward Gilbert Abbott. This occurred in the surgical amphitheater now called the Ether Dome. The previously skeptical Warren was impressed and stated, "Gentlemen, this is no humbug." In a letter to Morton shortly thereafter, physician and writer Oliver Wendell Holmes, Sr. proposed naming the state produced "anesthesia", and the procedure an "anesthetic".

Morton at first attempted to hide the actual nature of his anesthetic substance, referring to it as Letheon. He received a US patent for his substance, but news of the successful anesthetic spread quickly by late 1846. Respected surgeons in Europe including Liston, Dieffenbach, Pirogov, and Syme quickly undertook numerous operations with ether. An American-born physician, Boott, encouraged London dentist James Robinson to perform a dental procedure on a Miss Lonsdale. This was the first case of an operator-anesthetist. On the same day, 19 December 1846, in Dumfries Royal Infirmary, Scotland, a Dr. Scott used ether for a surgical procedure. The first use of anesthesia in the Southern Hemisphere took place in Launceston, Tasmania, that same year. Drawbacks with ether such as excessive vomiting and its explosive flammability led to its replacement in England with chloroform.

Discovered in 1831 by an American physician Samuel Guthrie (1782–1848), and independently a few months later by Frenchman Eugène Soubeiran (1797–1859) and Justus von Liebig (1803–73) in Germany, chloroform was named and chemically characterised in 1834 by Jean-Baptiste Dumas (1800–84). In 1842, Dr Robert Mortimer Glover in London discovered the anaesthetic qualities of chloroform on laboratory animals. In 1847, Scottish obstetrician James Young Simpson was the first to demonstrate the anesthetic properties of chloroform on humans and helped to popularise the drug for use in medicine. Its use spread quickly and gained royal approval in 1853 when John Snowgave it to Queen Victoria during the birth of Prince Leopold. During the birth itself, chloroform met all the Queen's expectations; she stated it was "delightful beyond measure". Unfortunately, though free of ether's flammability and consequent explosion hazard, chloroform is not as safe pharmacologically, especially when administered by an untrained practitioner (medical students, nurses, and occasionally members of the public were often pressed into giving anesthetics at this time). This led to many deaths from the use of chloroform that (with hindsight) might have been preventable. The first fatality directly attributed to chloroform anesthesia was recorded on 28 January 1848 after the death of Hannah Greener.

John Snow of London published articles from May 1848 onwards "On Narcotism by the Inhalation of Vapours" in the London Medical Gazette. Snow also involved himself in the production of equipment needed for the administration of inhalational anesthetics, the forerunner of today's anesthesia machines.

The first comprehensive medical textbook on the subject, Anesthesia, was authored in 1914 by anesthesiologist Dr. James Tayloe Gwathmey and the chemist Dr. Charles Baskerville. This book served as the standard reference for the specialty for decades and included details on the history of anesthesia as well as the physiology and techniques of inhalation, rectal, intravenous, and spinal anesthesia.

Of these first famous anesthetics, only nitrous oxide is still widely used today, with chloroform and ether having been replaced by safer but sometimes more expensive general anesthetics, and cocaine by more effective local anesthetics with less abuse potential.

Society and culture

Almost all healthcare providers use anesthetic drugs to some degree, but most health professions have their own field of specialists in the field including medicine, nursing and dentistry.

Doctors specializing in anaesthesiology, including perioperative care, development of an anesthetic plan, and the administration of anesthetics are known in the US as anesthesiologists and in the UK, Canada, Australia, and NZ as anaesthetists or anaesthesiologists. All anesthetics in the UK, Australia, New Zealand, Hong Kong and Japan are administered by doctors. Nurse anesthetists also administer anesthesia in 109 nations. In the US, 35% of anesthetics are provided by physicians in solo practice, about 55% are provided by anesthesia care teams (ACTs) with anesthesiologists medically directing anesthesiologist assistants or certified registered nurse anesthetists (CRNAs), and about 10% are provided by CRNAs in solo practice. There can also be anesthesiologist assistants (US) or physicians' assistants (anaesthesia) (UK) who assist with anesthesia.

Special populations

There are many circumstances when anesthesia needs to be altered for special circumstances due to the procedure (such as in cardiac surgery, cardiothoracic anesthesiology or neurosurgery), the patient (such as in pediatric anesthesia, geriatric, bariatric or obstetrical anesthesia) or special circumstances (such as in trauma, prehospital care, robotic surgeryor extreme environments).

Antipyreti

Coated 200 m g tablets of ibuprofen, a common antipyretic

Antipyretics are substances that reduce fever. Antipyretics cause the hypothalamus to override a prostaglandin-induced increase in temperature. The body then works to lower the temperature, which results in a reduction in fever.

Most antipyretic medications have other purposes. The most common antipyretics in the United States are ibuprofen and aspirin, which are nonsteroidal anti-inflammatory drugs (NSAIDs) used primarily as analgesics (pain relievers), but which also have antipyretic properties; and acetaminophen (paracetamol), an analgesic with weak anti-inflammatory properties.

There is some debate over the appropriate use of such medications, as fever is part of the body's immune response to infection. A study published by the Royal Society claims fever suppression causes at least 1% more influenza cases of death in the United States, which results in at least 700 extra deaths per year.

Bathing or sponging with lukewarm or cool water can effectively reduce body temperature in those with heat illness, but not usually in those with fever. The use of alcohol baths is not an appropriate cooling method, because there have been reported adverse events associated with systemic absorption of alcohol.

Medications

Many medications have antipyretic effects and thus are useful for fever but not in treating illness, including:

- NSAIDs such as ibuprofen, naproxen, ketoprofen, and nimesulide
- Aspirin, and related salicylates such as choline salicylate, magnesium salicylate, and sodium salicylate
- Paracetamol (USAN and JAN: acetaminophen)
- Metamizole, banned in over 30 countries for causing agranulocytosis
- Nabumetone
- Phenazone (antipyrine), available in combination with benzocaine as an ear drop in the US.

Children

The U.S. Food and Drug Administration (FDA) notes that improper dosing is one of the biggest problems in giving acetaminophen (paracetamol) to children. The effectiveness of acetaminophen alone as an antipyretic in children is uncertain, with some evidence showing it is no better than physical methods. Therapies involving alternating doses of acetaminophen and ibuprofen have shown greater antipyretic effect than either drug alone. One meta-analysis indicated that ibuprofen is more effective than acetaminophen in children at similar doses when both are given alone.

Due to concerns about Reye syndrome, it is recommend that aspirin and combination products containing aspirin not be given to children or teenagers during episodes of fever-causing illnesses.

Plants

Traditional use of higher plants with antipyretic properties is a common worldwide feature of many ethnobotanical cultural systems. In ethnobotany, plants with naturally occurring antipyretic properties are commonly referred to as febrifuge.

Popular culture

Antipyretic was the word spelled by Joanne Lagatta to win the 1991 Scripps National Spelling Bee.

On the second disc for the Final Fantasy Tactics soundtrack, there is a track titled Antipyretic.

Aspirin

Aspirin, also known as acetylsalicylic acid (ASA), is a medication used to treat pain, fever, or inflammation. Specific inflammatory conditions which aspirin is used to treat include Kawasaki disease, pericarditis, and rheumatic fever. Aspirin given shortly after a heart attack decreases the risk of death. Aspirin is also used long-term to help prevent further heart attacks, ischaemic strokes, and blood clots in people at high risk. It may also decrease the risk of certain types of cancer, particularly colorectal cancer. For pain or fever, effects typically begin within 30 minutes. Aspirin is a nonsteroidal anti-inflammatory drug (NSAID) and works similarly to other NSAIDs but also suppresses the normal functioning of platelets.

One common adverse effect is an upset stomach. More significant side effects include stomach ulcers, stomach bleeding, and worsening asthma. Bleeding risk is greater among those who are older, drink alcohol, take other NSAIDs, or are on other blood thinners. Aspirin is not recommended in the last part of pregnancy. It is not generally recommended in children with infectionsbecause of the risk of Reye syndrome. High doses may result in ringing in the ears.

A precursor to aspirin found in leaves from the willow tree has been used for its health effects for at least 2,400 years. In 1853, chemist Charles Frédéric Gerhardt treated the medicine sodium salicylate with acetyl chloride to produce acetylsalicylic acid for the first time. For the next fifty years, other chemists established the chemical structure and came up with more efficient methods to make it. In 1897, scientists at the Bayer company began studying acetylsalicylic acid as a less-irritating replacement medication for common salicylate medicines. By 1899, Bayer had named it "Aspirin" and sold it around the world. Aspirin's popularity grew over the first half of the twentieth century leading to competition between many brands and formulations. The word Aspirin was Bayer's brand name; however, their rights to the trademark were lost or sold in many countries.

Aspirin is one of the most widely used medications globally, with an estimated 40,000 tonnes (44,000 tons) (50 to 120 billion pills) consumed each year. It is on the World Health Organization's (WHO's) List of Essential Medicines, the safest and most effective medicines needed in a health system. As of 2014 the wholesale cost in the developing world is $0.002 to $0.025 USD per dose.As of 2015 the cost for a typical month of medication in the United States is less than US$25.00. It is available as a generic medication.

MEDICAL USE

Aspirin is used in the treatment of a number of conditions, including fever, pain, rheumatic fever, and inflammatory diseases, such as rheumatoid arthritis, pericarditis, and Kawasaki disease. Lower doses of aspirin have also been shown to reduce the risk of death from a heart attack, or the risk of stroke in people who are at high risk or who have cardiovascular disease, but not in elderly people who are otherwise healthy. There is some evidence that aspirin is effective at preventing colorectal cancer, though the mechanisms of this effect are unclear. In the United States, low dose aspirin is deemed reasonable in those between 50 and 70 years old who have a more than 10% risk of cardiovascular disease, and are not at an increased risk of bleeding, and who are otherwise healthy.

Pain

Aspirin is an effective analgesic for acute pain, but is generally considered inferior to ibuprofen for the alleviation of pain because aspirin is more likely to cause gastrointestinal bleeding. Aspirin is generally ineffective for those pains caused by muscle cramps, bloating, gastric distension, or acute skin irritation. As with other NSAIDs, combinations of aspirin and caffeine provide slightly greater pain relief than aspirin alone. Effervescent formulations of aspirin relieve pain faster than aspirin in tablets, which makes them useful for the treatment of migraines. Topical aspirin may be effective for treating some types of neuropathic pain.

Headache

Aspirin, either by itself or in a combined formulation, effectively treats certain types of a headache, but its efficacy may be questionable for others. Secondary headaches, meaning those caused by another disorder or trauma, should be promptly treated by a medical provider. Among primary headaches, the International Classification of Headache Disorders distinguishes between tension headache (the most common), migraine, and cluster headache. Aspirin or other over-the-counter analgesics are widely recognized as effective for the treatment of tension headache.

Aspirin, especially as a component of an aspirin/paracetamol/caffeine combination, is considered a first-line therapy in the treatment of migraine, and comparable to lower doses of sumatriptan. It is most effective at stopping migraines when they are first beginning.

Fever

Like its ability to control pain, aspirin's ability to control fever is due to its action on the prostaglandin system through its irreversible inhibition of COX. Although aspirin's use as an antipyretic in adults is well established, many medical societies and regulatory agencies (including the American Academy of Family Physicians, the American Academy of Pediatrics, and the U.S. Food and Drug Administration (FDA)) strongly advise against using aspirin for treatment of fever in children because of the risk of Reye's syndrome, a rare but often fatal illness associated with the use of aspirin or other salicylates in children during episodes of viral or bacterial infection. Because of the risk of Reye's syndrome in children, in 1986, the FDA required labeling on all aspirin-containing medications advising against its use in children and teenagers.

Inflammation

Aspirin is used as an anti-inflammatory agent for both acute and long-term inflammation, as well as for treatment of inflammatory diseases, such as rheumatoid arthritis.

Heart attacks and strokes

Aspirin is an important part of the treatment of those who have had a myocardial infarction (heart attack).

High risk

For people who have already had a heart attack or stroke, taking aspirin daily for two years prevented 1 in 50 from having a cardiovascular problem (heart attack, stroke, or death), but also caused non-fatal bleeding problems to occur in 1 of 400 people. Low dose aspirin appears useful for people less than 70 kg while higher dose aspirin is required to benefit those over 70 kg.

Lower risk

In those with no previous history of heart disease, aspirin decreases the risk of a non-fatal myocardial infarction but increases the risk of bleeding and does not change the overall risk of death. Specifically over 5 years it decreased the risk of a cardiovascular event by 1 in 265 and increased the risk of bleeding by 1 in 210.

Aspirin appears to offer little benefit to those at lower risk of heart attack or stroke—for instance, those without a history of these events or with pre-existing disease. Some studies recommend aspirin on a case-by-case basis, while others have suggested the risks of other events, such as gastrointestinal bleeding, were enough to outweigh any potential benefit, and recommended against using aspirin for primary prevention entirely. Aspirin has also been suggested as a component of a polypill for prevention of cardiovascular disease.

Complicating the use of aspirin for prevention is the phenomenon of aspirin resistance. For people who are resistant, aspirin's efficacy is reduced. Some authors have suggested testing regimens to identify people who are resistant to aspirin.

After surgery

After percutaneous coronary interventions (PCIs), such as the placement of a coronary artery stent, a U.S. Agency for Healthcare Research and Quality guideline recommends that aspirin be taken indefinitely. Frequently, aspirin is combined with an ADP receptor inhibitor, such as clopidogrel, prasugrel, or ticagrelor to prevent blood clots. This is called dual antiplatelet therapy (DAPT). United States and European Union guidelines disagree somewhat about how long, and for what indications this combined therapy should be continued after surgery. U.S. guidelines recommend DAPT for at least 12 months, while EU guidelines recommend DAPT for 6–12 months after a drug-eluting stent placement. However, they agree that aspirin be continued indefinitely after DAPT is complete.

Cancer prevention

Aspirin is thought to reduce the overall risk of both getting cancer and dying from cancer. This effect is particularly beneficial for colorectal cancer (CRC) but must be taken for at least 10–20 years to see this benefit. It may also slightly reduce the risk of endometrial cancer, breast cancer, and prostate cancer.

Some conclude the benefits are greater than the risks due to bleeding in those at average risk. Others are unclear if the benefits are greater than the risk. Given this uncertainty, the 2007 United States Preventive Services Task Force guidelines on this topic recommended against the use of aspirin for prevention of CRC in people with average risk. Nine years later however, the USPSTF issued a grade B recommendation for the use of low-dose aspirin (75 to 100 mg/day) "for the primary prevention of CVD [cardiovascular disease] and CRC in adults 50 to 59 years of age who have a 10% or greater 10-year CVD risk, are not at increased risk for bleeding, have a life expectancy of at least 10 years, and are willing to take low-dose aspirin daily for at least 10 years".

Other uses

Aspirin is a first-line treatment for the fever and joint-pain symptoms of acute rheumatic fever. The therapy often lasts for one to two weeks, and is rarely indicated for longer periods. After fever and pain have subsided, the aspirin is no longer necessary, since it does not decrease the incidence of heart complications and residual rheumatic heart disease. Naproxen has been shown to be as effective as aspirin and less toxic, but due to the limited clinical experience, naproxen is recommended only as a second-line treatment. Along with rheumatic fever, Kawasaki disease remains one of the few indications for aspirin use in children in spite of a lack of high quality evidence for its effectiveness.

Low-dose aspirin supplementation has moderate benefits when used for prevention of pre-eclampsia. This benefit is greater when started in early pregnancy. There is no evidence that aspirin prevents dementia.

Resistance

For some people, aspirin does not have as strong an effect on platelets as for others, an effect known as aspirin-resistance or insensitivity. One study has suggested women are more likely to be resistant than men, and a different, aggregate study of 2,930 people found 28% were resistant. A study in 100 Italian people, though, found, of the apparent 31% aspirin-resistant subjects, only 5% were truly resistant, and the others were noncompliant. Another study of 400 healthy volunteers found no subjects who were truly resistant, but some had "pseudoresistance, reflecting delayed and reduced drug absorption".

Dosage

Adult aspirin tablets are produced in standardised sizes, which vary slightly from country to country, for example 300 mg in Britain and 325 mg (or 5 grains) in the United States. Smaller doses are based on these standards, e.g., 75 mg and 81 mg tablets. The 81 mg (1 D 4-grain) tablets are commonly called "baby aspirin" or "baby-strength", because they were originally—but no longer—intended to be administered to infants and children. No medical significance occurs due to the slight difference in dosage between the 75 mg and the 81 mg tablets. The dose required for benefit appears to depend on a person's weight. For those less than 70 kg low dose is effective for preventing cardiovascular disease while in those who weight more than 70 kg higher doses are required.

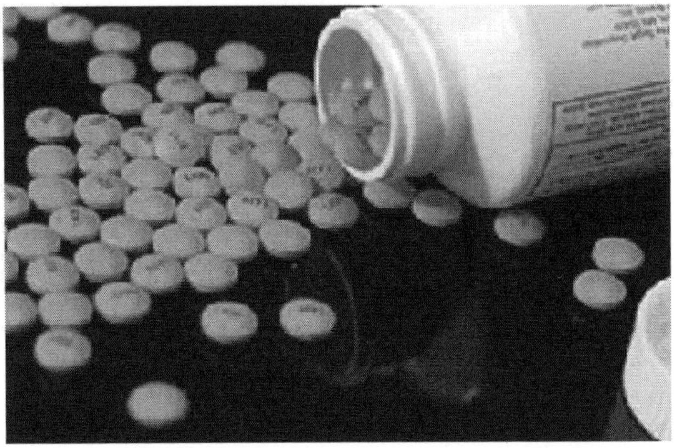

Coated 325 mg (5-grain) aspirin tablets

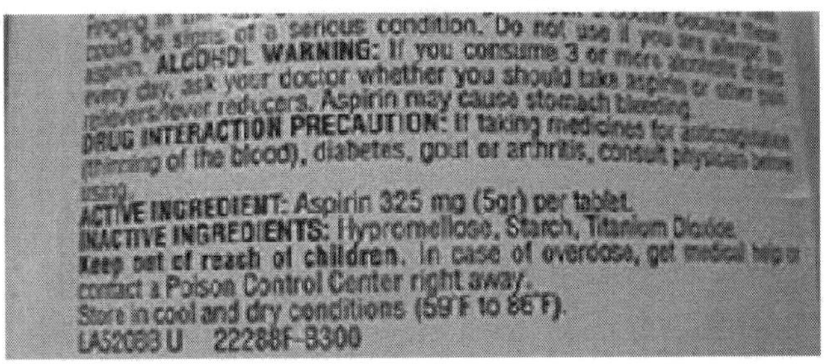

The 5-grain aspirin. The usage guidance label on a bottle of aspirin indicates that the dosage is "325 mg (5 gr)".

In general, for adults, doses are taken four times a day for fever or arthritis, with doses near the maximal daily dose used historically for the treatment of rheumatic fever. For the prevention of myocardial infarction (MI) in someone with documented or suspected coronary artery disease, much lower doses are taken once daily.

March 2009 recommendations from the USPSTF on the use of aspirin for the primary prevention of coronary heart disease encourage men aged 45–79 and women aged 55–79 to use aspirin when the potential benefit of a reduction in MI for men or stroke for women outweighs the potential harm of an increase in gastrointestinal hemorrhage. The WHI study said regular low dose (75 or 81 mg) aspirin female users had a 25% lower risk of death from cardiovascular disease and a 14% lower risk of death from any cause. Low-dose aspirin use was also associated with a trend toward lower risk of cardiovascular events, and lower aspirin doses (75 or 81 mg/day) may optimize efficacy and safety for people requiring aspirin for long-term prevention.

In children with Kawasaki disease, aspirin is taken at dosages based on body weight, initially four times a day for up to two weeks and then at a lower dose once daily for a further six to eight weeks.

Adverse effects

Contraindications

Aspirin should not be taken by people who are allergic to ibuprofen or naproxen, or who have salicylate intolerance or a more generalized drug intolerance to NSAIDs, and caution should be exercised in those with asthma or NSAID-precipitated bronchospasm. Owing to its effect on the stomach lining, manufacturers recommend people with peptic ulcers, mild diabetes, or gastritis seek medical advice before using aspirin. Even if none of these conditions is present, the risk of stomach bleeding is still increased when aspirin is taken with alcohol or warfarin. People with hemophilia or other bleeding tendencies should not take aspirin or other salicylates. Aspirin is known to cause hemolytic anemia in people who have the genetic disease glucose-6-phosphate dehydrogenase deficiency, particularly in large doses and depending on the severity of the disease. Use of aspirin during dengue fever is not recommended owing to increased bleeding tendency. People with kidney disease, hyperuricemia, or gout should not take aspirin because it inhibits the kidneys' ability to excrete uric acid, thus may exacerbate these conditions. Aspirin should not be given to children or adolescents to control cold or influenza symptoms, as this has been linked with Reye's syndrome.

Gastrointestinal

Aspirin use has been shown to increase the risk of gastrointestinal bleeding. Although some enteric-coated formulations of aspirin are advertised as being "gentle to the stomach", in one study, enteric coating did not seem to reduce this risk. Combining aspirin with other NSAIDs has also been shown to further increase this risk. Using aspirin in combination with clopidogrel or warfarin also increases the risk of upper gastrointestinal bleeding.

Blockade of COX-1 by aspirin apparently results in the upregulation of COX-2 as part of a gastric defense and that taking COX-2 inhibitors concurrently with aspirin increases the gastric mucosal erosion. Therefore, caution should be exercised if combining aspirin with any "natural" supplements with COX-2-inhibiting properties, such as garlic extracts, curcumin, bilberry, pine bark, ginkgo, fish oil, resveratrol, genistein, quercetin, resorcinol, and others.

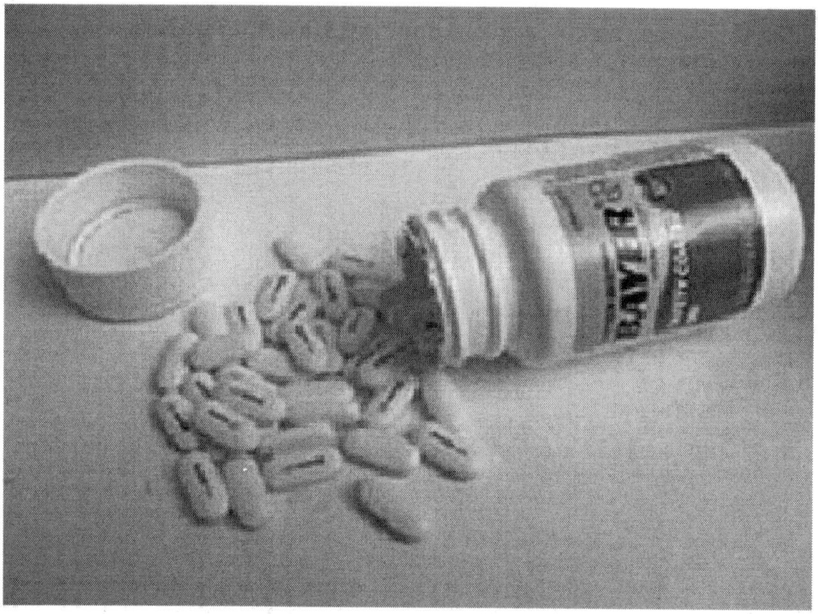

Enteric-coated 325 mg aspirin pills

In addition to enteric coating, "buffering" is the other main method companies have used to try to mitigate the problem of gastrointestinal bleeding. Buffering agents are intended to work by preventing the aspirin from concentrating in the walls of the stomach, although the benefits of buffered aspirin are disputed. Almost any buffering agent used in antacids can be used; Bufferin, for example, uses magnesium oxide. Other preparations use calcium carbonate.

Taking it with vitamin C has been investigated as a method of protecting the stomach lining. Taking equal doses of vitamin C and aspirin may decrease the amount of stomach damage that occurs compared to taking aspirin alone.

Central effects

Large doses of salicylate, a metabolite of aspirin, cause temporary tinnitus (ringing in the ears) based on experiments in rats, via the action on arachidonic acid and NMDA receptorscascade.

Reye's syndrome

Reye's syndrome, a rare but severe illness characterized by acute encephalopathy and fatty liver, can occur when children or adolescents are given aspirin for a fever or other illness or infection. From 1981 through 1997, 1207 cases of Reye's syndrome in people younger than 18 were reported to the U.S. Centers for Disease Control and Prevention. Of these, 93% reported being ill in the three weeks preceding the onset of Reye's syndrome, most commonly with a respiratory infection, chickenpox, or diarrhea. Salicylates were detectable in 81.9% of children for whom test results were reported. After the association between Reye's syndrome and aspirin was reported, and safety measures to prevent it (including a Surgeon General's warning, and changes to the labeling of aspirin-containing drugs) were implemented, aspirin taken by children declined considerably in the United States, as did the number of reported cases of Reye's syndrome; a similar decline was found in the United Kingdom after warnings against pediatric aspirin use were issued. The U.S. Food and Drug Administration now recommends aspirin (or aspirin-containing products) should not be given to anyone under the age of 12 who has a fever, and the UK National Health Service recommends children who are under 16 years of age should not take aspirin, unless it is on the advice of a doctor.

Skin

For a small number of people, taking aspirin can result in symptoms resembling an allergic reaction, including hives, swelling, and headache. The reaction is caused by salicylate intolerance and is not a true allergy, but rather an inability to metabolize even small amounts of aspirin, resulting in an overdose.

Aspirin and other NSAIDs, such as ibuprofen, may delay the healing of skin wounds. Aspirin may however help heal venous leg ulcers that have not healed following usual treatment.

OTHER ADVERSE EFFECTS

Aspirin can induce swelling of skin tissues in some people. In one study, angioedema appeared one to six hours after ingesting aspirin in some of the people. However, when the aspirin was taken alone, it did not cause angioedema in these people; the aspirin had been taken in combination with another NSAID-induced drug when angioedema appeared.

Aspirin causes an increased risk of cerebral microbleeds having the appearance on MRI scans of 5 to 10 mm or smaller, hypointense (dark holes) patches. Such cerebral microbleeds are important, since they often occur prior to ischemic stroke or intracerebral hemorrhage, Binswanger disease, and Alzheimer's disease.

A study of a group with a mean dosage of aspirin of 270 mg per day estimated an average absolute risk increase in intracerebral hemorrhage (ICH) of 12 events per 10,000 persons. In comparison, the estimated absolute risk reduction in myocardial infarction was 137 events per 10,000 persons, and a reduction of 39 events per 10,000 persons in ischemic stroke. In cases where ICH already has occurred, aspirin use results in higher mortality, with a dose of about 250 mg per day resulting in a relative risk of death within three months after the ICH around 2.5 (95% confidence interval 1.3 to 4.6).

Aspirin and other NSAIDs can cause abnormally high blood levels of potassium by inducing a hyporeninemic hypoaldosteronic state via inhibition of prostaglandin synthesis; however, these agents do not typically cause hyperkalemia by themselves in the setting of normal renal function and euvolemic state.

Aspirin can cause prolonged bleeding after operations for up to 10 days. In one study, 30 of 6499 people having elective surgery required reoperations to control bleeding. Twenty had diffuse bleeding and 10 had bleeding from a site. Diffuse, but not discrete, bleeding was associated with the preoperative use of aspirin alone or in combination with other NSAIDS in 19 of the 20 diffuse bleeding people.

On 9 July 2015, the FDA toughened warnings of increased heart attack and stroke risk associated with nonsteroidal anti-inflammatory drugs (NSAID). Aspirin is an NSAID but is not affected by the new warnings.

Overdose

Aspirin overdose can be acute or chronic. In acute poisoning, a single large dose is taken; in chronic poisoning, higher than normal doses are taken over a period of time. Acute overdose has a mortality rate of 2%. Chronic overdose is more commonly lethal, with a mortality rate of 25%; chronic overdose may be especially severe in children. Toxicity is managed with a number of potential treatments, including activated charcoal, intravenous dextrose and normal saline, sodium bicarbonate, and dialysis. The diagnosis of poisoning usually involves measurement of plasma salicylate, the active metabolite of aspirin, by automated spectrophotometric methods. Plasma salicylate levels in general range from 30–100 mg/l after usual therapeutic doses, 50–300 mg/l in people taking high doses and 700–1400 mg/l following acute overdose. Salicylate is also produced as a result of exposure to bismuth subsalicylate, methyl salicylate, and sodium salicylate.

Interactions

Aspirin is known to interact with other drugs. For example, acetazolamide and ammonium chloride are known to enhance the intoxicating effect of salicylates, and alcohol also increases the gastrointestinal bleeding associated with these types of drugs. Aspirin is known to displace a number of drugs from protein-binding sites in the blood, including the antidiabetic drugs tolbutamide and chlorpropamide, warfarin, methotrexate, phenytoin, probenecid, valproic acid (as well as interfering with beta oxidation, an important part of valproate metabolism), and other NSAIDs. Corticosteroids may also reduce the concentration of aspirin. Ibuprofen can negate the antiplatelet effect of aspirin used for cardioprotection and stroke prevention. The pharmacological activity of spironolactone may be reduced by taking aspirin, and it is known to compete with penicillin G for renal tubular secretion. Aspirin may also inhibit the absorption of vitamin C.

Chemical properties

Aspirin decomposes rapidly in solutions of ammonium acetate or the acetates, carbonates, citrates, or hydroxides of the alkali metals. It is stable in dry air, but gradually hydrolyses in contact with moisture to acetic and salicylic acids. In solution with alkalis, the hydrolysis proceeds rapidly and the clear solutions formed may consist entirely of acetate and salicylate.

Like flour mills, factories that make aspirin tablets must pay attention to how much of the powder gets into the air inside the building, because the powder-air mixture can be explosive. The National Institute for Occupational Safety and Health (NIOSH) has set a recommended exposure limit in the United States of 5 mg/m (time-weighted average). In 1989, the Occupational Safety and Health Administration (OSHA) set a legal permissible exposure limit for aspirin of 5 mg/m, but this was vacated by the AFL-CIO v. OSHA decision in 1993.

Polymorphism

Polymorphism, or the ability of a substance to form more than one crystal structure, is important in the development of pharmaceutical ingredients. Many drugs are receiving regulatory approval for only a single crystal form or polymorph. For a long time, only one crystal structure for aspirin was known. That aspirin might have a second crystalline form was suspected since the 1960s. The elusive second polymorph was first discovered by Vishweshwar and coworkers in 2005, and fine structural details were given by Bond et al. A new crystal type was found after attempted cocrystallization of aspirin and levetiracetam from hot acetonitrile. The form II is only stable at 100 K and reverts to form I at ambient temperature. In the (unambiguous) form I, two salicylic molecules form centrosymmetric dimers through the acetyl groups with the (acidic) methyl proton to carbonylhydrogen bonds, and in the newly claimed form II, each salicylic molecule forms the same hydrogen bonds with two neighboring molecules instead of one. With respect to the hydrogen bonds formed by the carboxylic acid groups, both polymorphs form identical dimer structures.

MECHANISM OF ACTION

Discovery of the mechanism

In 1971, British pharmacologist John Robert Vane, then employed by the Royal College of Surgeons in London, showed aspirin suppressed the production of prostaglandins and thromboxanes. For this discovery he was awarded the 1982 Nobel Prize in Physiology or Medicine, jointly with Sune Bergström and Bengt Ingemar Samuelsson.

Prostaglandins and thromboxanes

Aspirin's ability to suppress the production of prostaglandins and thromboxanes is due to its irreversible inactivation of the cyclooxygenase (COX; officially known as prostaglandin-endoperoxide synthase, PTGS) enzyme required for prostaglandin and thromboxane synthesis. Aspirin acts as an acetylating agent where an acetyl group is covalently attached to a serine residue in the active site of the PTGS enzyme (Suicide inhibition). This makes aspirin different from other NSAIDs (such as diclofenac and ibuprofen), which are reversible inhibitors.

Low-dose aspirin use irreversibly blocks the formation of thromboxane A2 in platelets, producing an inhibitory effect on platelet aggregation during the lifetime of the affected platelet (8–9 days). This antithrombotic property makes aspirin useful for reducing the incidence of heart attacks in people who have had a heart attack, unstable angina, ischemic stroke or transient ischemic attack. 40 mg of aspirin a day is able to inhibit a large proportion of maximum thromboxane A2 release provoked acutely, with the prostaglandin I2 synthesis being little affected; however, higher doses of aspirin are required to attain further inhibition.

Prostaglandins, local hormones produced in the body, have diverse effects, including the transmission of pain information to the brain, modulation of the hypothalamic thermostat, and inflammation. Thromboxanes are responsible for the aggregation of platelets that form blood clots. Heart attacks are caused primarily by blood clots, and low doses of aspirin are seen as an effective medical intervention for acute myocardial infarction.

COX-1 and COX-2 inhibition

At least two different types of cyclooxygenases, COX-1 and COX-2, are acted on by aspirin. Aspirin irreversibly inhibits COX-1 and modifies the enzymatic activity of COX-2. COX-2 normally produces prostanoids, most of which are proinflammatory. Aspirin-modified PTGS2 produces lipoxins, most of which are anti-inflammatory. Newer NSAID drugs, COX-2 inhibitors (coxibs), have been developed to inhibit only PTGS2, with the intent to reduce the incidence of gastrointestinal side effects.

However, several of the new COX-2 inhibitors, such as rofecoxib (Vioxx), have been withdrawn in the last decade, after evidence emerged that PTGS2 inhibitors increase the risk of heart attack and stroke. Endothelial cells lining the microvasculature in the body are proposed to express PTGS2, and, by selectively inhibiting PTGS2, prostaglandin production (specifically, PGI2; prostacyclin) is downregulated with respect to thromboxane levels, as PTGS1 in platelets is unaffected. Thus, the protective anticoagulative effect of PGI2 is removed, increasing the risk of thrombus and associated heart attacks and other circulatory problems. Since platelets have no DNA, they are unable to synthesize new PTGS once aspirin has irreversibly inhibited the enzyme, an important difference with reversible inhibitors.

Furthermore, aspirin, while inhibiting the ability of COX-2 to form pro-inflammatory products such as the prostaglandins, converts this enzyme's activity from a prostaglandin-forming cyclooxygenase to a lipoxygenase-like enzyme: aspirin-treated COX-2 metabolizes a variety of polyunsaturated fatty acids to hydroperoxy products which are then further metabolized to specialized proresolving mediators such as the aspirin-triggered lipoxins, aspirin-triggered resolvins, and aspirin-triggered maresins. These mediators possess potent anti-inflammatory activity. It is proposed that this aspirin-triggered transition of COX-2 from cyclooxygenase to lipoxygenase activity and the consequential formation of specialized proresolving mediators contributes to the anti-inflammatory effects of aspirin.

Additional mechanisms

Aspirin has been shown to have at least three additional modes of action. It uncouples oxidative phosphorylation in cartilaginous (and hepatic) mitochondria, by diffusing from the inner membrane space as a proton carrier back into the mitochondrial matrix, where it ionizes once again to release protons. Aspirin buffers and transports the protons. When high doses are given, it may actually cause fever, owing to the heat released from the electron transport chain, as opposed to the antipyretic action of aspirin seen with lower doses. In addition, aspirin induces the formation of NO-radicals in the body, which have been shown in mice to have an independent mechanism of reducing inflammation. This reduced leukocyte adhesion is an important step in the immune response to infection; however, evidence is insufficient to show aspirin helps to fight infection. More recent data also suggest salicylic acid and its derivatives modulate signaling through NF-êB.

NF-êB, a transcription factor complex, plays a central role in many biological processes, including inflammation.

Aspirin is readily broken down in the body to salicylic acid, which itself has anti-inflammatory, antipyretic, and analgesic effects. In 2012, salicylic acid was found to activate AMP-activated protein kinase, which has been suggested as a possible explanation for some of the effects of both salicylic acid and aspirin. The acetyl portion of the aspirin molecule has its own targets. Acetylation of cellular proteins is a well-established phenomenon in the regulation of protein function at the post-translational level. Aspirin is able to acetylate several other targets in addition to COX isoenzymes. These acetylation reactions may explain many hitherto unexplained effects of aspirin.

Pharmacokinetics

Acetylsalicylic acid is a weak acid, and very little of it is ionized in the stomach after oral administration. Acetylsalicylic acid is quickly absorbed through the cell membrane in the acidic conditions of the stomach. The increased pH and larger surface area of the small intestine causes aspirin to be absorbed more slowly there, as more of it is ionised. Owing to the formation of concretions, aspirin is absorbed much more slowly during overdose, and plasma concentrations can continue to rise for up to 24 hours after ingestion.

About 50–80% of salicylate in the blood is bound to albumin protein, while the rest remains in the active, ionized state; protein binding is concentration-dependent. Saturation of binding sites leads to more free salicylate and increased toxicity. The volume of distribution is 0.1–0.2 L/kg. Acidosis increases the volume of distribution because of enhancement of tissue penetration of salicylates.

As much as 80% of therapeutic doses of salicylic acid is metabolized in the liver. Conjugation with glycine forms salicyluric acid, and with glucuronic acid to form two different glucuronide esters. The conjugate with the acetyl group intact is referred to as the acyl glucuronide; the deacetylated conjugate is the phenolic glucuronide. These metabolic pathways have only a limited capacity. Small amounts of salicylic acid are also hydroxylated to gentisic acid. With large salicylate doses, the kinetics switch from first-order to zero-order, as metabolic pathways become saturated and renal excretion becomes increasingly important.

Salicylates are excreted mainly by the kidneys as salicyluric acid (75%), free salicylic acid (10%), salicylic phenol (10%), and acyl glucuronides (5%), gentisic acid (< 1%), and 2,3-dihydroxybenzoic acid. When small doses (less than 250 mg in an adult) are ingested, all pathways proceed by first-order kinetics, with an elimination half-life of about 2.0 h to 4.5 h. When higher doses of salicylate are ingested (more than 4 g), the half-life becomes much longer (15 h to 30 h), because the biotransformation pathways concerned with the formation of salicyluric acid and salicyl phenolic glucuronide become saturated. Renal excretion of salicylic acid becomes increasingly important as the metabolic pathways become saturated, because it is extremely sensitive to changes in urinary pH. A 10- to 20-fold increase in renal clearance occurs when urine pH is increased from 5 to 8. The use of urinary alkalinization exploits this particular aspect of salicylate elimination. It was found that short-term aspirin use in therapeutic doses might precipitate reversible acute renal failure when the patient was ill with glomerulonephritis or cirrhosis. Aspirin for some patients with chronic renal insufficiency and some children with congestive heart failure was contraindicated.

HISTORY

Medicines made from willow and other salicylate-rich plants appear in clay tablets from ancient Sumer as well as the Ebers Papyrus from ancient Egypt. Hippocrates referred to their use of salicylic tea to reduce fevers around 400 BC, and were part of the pharmacopoeia of Western medicine in classical antiquity and the Middle Ages. Willow bark extract became recognized for its specific effects on fever, pain and inflammation in the mid-eighteenth century. By the nineteenth century pharmacists were experimenting with and prescribing a variety of chemicals related to salicylic acid, the active component of willow extract.

In 1853, chemist Charles Frédéric Gerhardt treated sodium salicylate with acetyl chloride to produce acetylsalicylic acid for the first time; in the second half of the nineteenth century, other academic chemists established the compound's chemical structure and devised more efficient methods of synthesis. In 1897, scientists at the drug and dye firm Bayer began investigating acetylsalicylic acid as a less-irritating replacement for standard common salicylate medicines, and identified a new way to synthesize it. By 1899, Bayer had dubbed this drug Aspirin and was selling it around the world. The word Aspirin was Bayer's brand name, rather than the generic name of the drug; however, Bayer's rights to the trademark were lost or sold in many countries. Aspirin's popularity grew over the first half of the twentieth century leading to fierce competition with the proliferation of aspirin brands and products.

Aspirin's popularity declined after the development of acetaminophen/ paracetamol in 1956 and ibuprofen in 1962. In the 1960s and 1970s, John Vane and others discovered the basic mechanism of aspirin's effects, while clinical trials and other studies from the 1960s to the 1980s established aspirin's efficacy as an anticlotting agent that reduces the risk of clotting diseases. The initial large studies on the use of low-dose aspirin to prevent heart attacks that were published in the 1970s and 1980s helped spur reform in clinical research ethics and guidelines for human subject research and US federal law, and are often cited as examples of clinical trials that included only men, but from which people drew general conclusions that did not hold true for women.

Aspirin sales revived considerably in the last decades of the twentieth century, and remain strong in the twenty-first with widespread use as a preventive treatment for heart attacksand strokes.

Trademark

Bayer lost its trademark for Aspirin in the United States in 1918 because it had failed to use the name for its own product and had for years allowed the use of "Aspirin" by other manufacturers . Today, aspirin is a generic trademark in many countries.Aspirin, with a capital "A", remains a registered trademark of Bayer in Germany, Canada, Mexico, and in over 80 other countries, for acetylsalicylic acid in all markets, but using different packaging and physical aspects for each.

Compendial status

- United States Pharmacopeia
- British Pharmacopoeia

Veterinary medicine

Aspirin is sometimes used in veterinary medicine as an anticoagulant or to relieve pain associated with musculoskeletal inflammation or osteoarthritis. Aspirin should only be given to animals under the direct supervision of a veterinarian, as adverse effects—including gastrointestinal issues—are common. An aspirin overdose in any species may result in salicylate poisoning, characterized by hemorrhaging, seizures, coma, and even death.

Cats and dogs

Dogs are better able to tolerate aspirin than cats are. Cats metabolize aspirin slowly because they lack the glucuronideconjugates that aid in the excretion of aspirin, making it potentially toxic if dosing is not spaced out properly. No clinical signs of toxicosis occurred when cats were given 25 mg/kg of aspirin every 48 hours for 4 weeks, but the recommended dose for relief of pain and fever and for treating blood clotting diseases in cats is 10 mg/kg every 48 hours to allow for metabolization.

United Nations

Single Convention on Narcotic Drugs, 1961

The adoption of this Convention is regarded as a milestone in the history of international drug ban. The Single Convention codified all existing multilateral treaties on drug control and extended the existing control systems to include the cultivation of plants that were grown as the raw material of narcotic drugs. The principal objectives of the Convention are to limit the possession, use, trade, distribution, import, export, manufacture and production of drugs exclusively to medical and scientific purposes, and to address drug trafficking through international cooperation to deter and discourage drug traffickers. The Convention also established the International Narcotics Control Board, merging the Permanent Central Board and the Drug Supervisory Board.

The 1961 Convention seeks to control more than 116 drugs that it classifies as narcotic. These include:

- Plant-based products such as opium and its derivatives morphine, codeine and heroin (the primary category of drug listed in the Convention);
- Synthetic narcotics such as methadone and pethidine; and
- Cannabis, coca and cocaine.

The Convention divides drugs into four groups, or schedules, in order to enforce a greater or lesser degree of control for the various substances and compounds. Opium smoking and eating, coca leaf chewing, cannabis resin smoking and the non-medical use of cannabis are prohibited. The 1972 Protocol to this Convention calls for increased efforts to prevent illicit production of, traffic in and use of narcotics as defined by the Convention, while highlighting the need to provide treatment and rehabilitation services to drug abusers.

INCB Yellow List

This document contains the current list of narcotic drugs under international control and additional information to assist governments in filling in the International Narcotics Control Board questionnaires related to narcotic drugs, namely, form A, form B and form C.

In medicine, a chemical agent that induces stupor, coma, or insensibility to pain (also called narcotic analgesic).

In the context of international drug control, "narcotic drug" means any drug defined as such under the 1961 Convention.

World Health Organization

Studies on the definition of counterfeit medicines in WHO member states

Assessment of the definitions of counterfeit medicines (or equivalent) in Member States

The nature of legal definitions: the unambiguity requirement

In order to avoid room for difference in interpretation, lawmakers (codificators) sometimes deviate from etymological (definiendum plus definientia) definitions. In doing so, they approach the term from the law enforcement point of view. The best example is the definition of narcotics in the United Nations Conventions. Narcotics are substances and preparations that induce drowsiness, sleep, stupor, insensibility, etc., and that these effects (and their rate) are complicated to prove, e.g. during litigation. Thus, the legal definition of a narcotic is whether or not it is listed on the Schedules of the Convention. If it is on some of the Schedules, it is narcotic.

Lexicon of alcohol and drug terms published by the World Health Organization

The term usually refers to opiates or opioids, which are called narcotic analgesics. In common parlance and legal usage, it is often used imprecisely to mean illicit drugs, irrespective of their pharmacology. For example, narcotics control legislation in Canada, USA, and certain other countries includes cocaine and cannabis as well as opioids (see also conventions, international drug). Because of this variation in usage, the term is best replaced by one with a more specific meaning (e.g. opioid).

United States

Section 1300.01 Definitions relating to controlled substances:

(b) As used in parts 1301 through 1308 and part 1312 of this chapter, the following terms shall have the meanings specified:

(30) The term narcotic drug means any of the following whether produced directly or indirectly by extraction from substances of vegetable origin or independently by means of chemical synthesis or by a combination of extraction and chemical synthesis:

(i) Opium, opiates, derivatives of opium and opiates, including their isomers, esters, ethers, salts, and salts of isomers, esters, and ethers whenever the existence of such isomers, esters, ethers and salts is possible within the specific chemical designation. Such term does not include the isoquinoline alkaloids of opium.

(ii) Poppy straw and concentrate of poppy straw.

(iii) Coca leaves, except coca leaves and extracts of coca leaves from which cocaine, ecgonine and derivatives of ecgonine or their salts have been removed.

(iv) Cocaine, its salts, optical and geometric isomers, and salts of isomers.

(v) Ecgonine, its derivatives, their salts, isomers and salts of isomers.

(vi) Any compound, mixture, or preparation which contains any quantity of any of the substances referred to in paragraphs (b)(31)(i) through (v) of this section.

A 1984 amendment to 21 USC (Controlled Substances Act), Section 802 expanded and revised definition of "narcotic drug", including within term poppy straw, cocaine, and ecgonine.

US v. Stieren

United States Court of Appeals, Eighth Circuit. Decided Oct. 31, 1979. LAY, Circuit Judge.

John Arthur Stieren appeals from the judgment of conviction for possession of cocaine with intent to distribute and dispense under 21 U.S.C. § 841(a)(1). Stieren contends that the statute is unconstitutional because "cocaine is classified as a narcotic under Schedule II of 21 U.S.C. § 812(c), when as a matter of scientific and medical fact cocaine is not a narcotic but is a non-narcotic stimulant."

The sufficiency of the evidence is not disputed. Stieren was convicted after special agents testified that he was in possession of and attempted to sell them a large quantity of cocaine. Defendant urges that the testimony and reports by physicians and scientists demonstrate that cocaine is not a narcotic. He also cites cases which hold that cocaine is not a narcotic under the pharmacological definition of the term. State v. Erickson, 574 P.2d 1 (Alaska 1978).

It is within the legislative prerogative to classify cocaine, which is a non-narcotic central nervous system stimulant, as a narcotic for penalty and regulatory purposes. 21 U.S.C. § 802(16)(A). The use of cocaine poses serious problems for the community and has a high potential for abuse. Congress' choice of penalty reflects a societal policy which must be adhered to by the courts.2 Congress has the power to reclassify cocaine. This power has been delegated to the Attorney General. 21 U.S.C. § 811(a)(1). If cocaine is to be reclassified, defendant's arguments should be made to the legislative branch, not the courts.

We hold that Congress had a rational legislative purpose when it classified cocaine as a Schedule II narcotic drug for the purpose of imposing penalties.

JUDGMENT AFFIRMED.

History

The term "narcotic" is believed to have been coined by the Greek physician Galen to refer to agents that numb or deaden, causing loss of feeling or paralysis. It is based on the Greek word íáñêùóéò (narcosis), the term used by Hippocrates for the process of numbing or the numbed state. Galen listed mandrake root, altercus (eclata),seeds, and poppy juice (opium) as the chief examples. It originally referred to any substance that relieved pain, dulled the senses, or induced sleep. Now, the term is used in a number of ways. Some people define narcotics as substances that bind at opioid receptors (cellular membrane proteins activated by substances like heroin or morphine) while others refer to any illicit substance as a narcotic. From a U.S. legal perspective, narcotics refer to opium, opium derivatives, and their semi-synthetic substitutes, though in U.S. law, due to its numbing properties, cocaine is also considered a narcotic.

Sense of "any illegal drug" first recorded 1926, Amer.Eng. The adj. is first attested c.1600. There are many different types of narcotics. The two most common forms of narcotic drugs are morphine and codeine. Both are synthesized from opium for medicinal use. The most commonly used drug for recreational purposes created from opium is heroin. Synthesized drugs created with an opium base for use in pain management are fentanyl, oxycodone, tramadol, demarol, hydrocodone, methadone, and hydromorphone. New forms of pain medication are being created regularly. The newest drug to come out in 2014 is zohydro, an intense dosage of hydrocodone medication, the strongest yet created for pain management.

Analgesics

Analgesics are drugs that relieve pain. There are two main types: non-narcotic analgesics for mild pain, and narcotic analgesics for severe pain.

Opioid

Opioids are substances that act on opioid receptors to produce morphine-like effects. Medically they are primarily used for pain relief, including anesthesia. Other medical uses include suppression of diarrhea, replacement therapy for opioid use disorder, reversing opioid overdose, suppressing cough, suppressing opioid induced constipation, as well as for executions in the United States. Extremely potent opioids such as carfentanil are only approved for veterinary use. Opioids are also frequently used non-medically for their euphoric effects or to prevent withdrawal.

Side effects of opioids may include itchiness, sedation, nausea, respiratory depression, constipation, and euphoria. Tolerance and dependence will develop with continuous use, requiring increasing doses and leading to a withdrawal syndrome upon abrupt discontinuation. The euphoria attracts recreational use and frequent, escalating recreational use of opioids typically results in addiction. An overdose or concurrent use with other depressant drugs commonly results in death from respiratory depression.

Opioids act by binding to opioid receptors, which are found principally in the central and peripheral nervous system and the gastrointestinal tract. These receptors mediate both the psychoactive and the somatic effects of opioids. Opioid drugs include partial agonists, like the anti-diarrhea drug loperamide and antagonists like naloxegol for opioid-induced constipation, which do not cross the blood-brain barrier, but can displace other opioids from binding to those receptors. Because opioids are addictive and may result in fatal overdose, most are controlled substances. In 2013, between 28 and 38 million people used opioids illicitly (0.6% to 0.8% of the global population between the ages of 15 and 65). In 2011, an estimated 4 million people in the United States used opioids recreationally or were dependent on them. As of 2015, increased rates of recreational use and addiction are attributed to over-prescription of opioid medications and inexpensive illicit heroin. Conversely, fears about over-prescribing, exaggerated side effects and addiction from opioids are similarly blamed for under-treatment of pain.

Terminology

Opioids include opiates, an older term that refers to such drugs derived from opium, including morphine itself. Other opioids are semi-synthetic and synthetic drugs such as hydrocodone, oxycodone and fentanyl; antagonist drugs such as naloxone; and endogenous peptides such as the endorphins. The terms opiate and narcotic are sometimes encountered as synonyms for opioid. Opiate is properly limited to the natural alkaloids found in the resin of the opium poppy although some include semi-synthetic derivatives. Narcotic, derived from words meaning 'numbness' or 'sleep', as an American legal term, refers to cocaine and opioids, and their source materials; it is also loosely applied to any illegal or controlled psychoactive drug. In some jurisdictions all controlled drugs are legally classified as narcotics. The term can have pejorative connotations and its use is generally discouraged where that is the case.

Medical uses

Pain

The weak opioid codeine, in low doses and combined with one or more other drugs, is commonly available without a prescription and can be used to treat mild pain. Other opioids are usually reserved for the relief of moderate to severe pain.

Acute pain

Opioids are effective for the treatment of acute pain (such as pain following surgery). For immediate relief of moderate to severe acute pain opioids are frequently the treatment of choice due to their rapid onset, efficacy and reduced risk of dependence. However a new report showed a clear risk of prolonged opioid use when opioid analgesics are initiated for an acute pain management following surgery or trauma. They have also been found to be important in palliative care to help with the severe, chronic, disabling pain that may occur in some terminal conditions such as cancer, and degenerative conditions such as rheumatoid arthritis. In many cases opioids are a successful long-term care strategy for those with chronic cancer pain.

Chronic non-cancer pain

Guidelines have suggested that the risk of opioids is likely greater than their benefits when used for most non-cancer chronic conditions including headaches, back pain, and fibromyalgia. Thus they should be used cautiously in chronic non-cancer pain. If used the benefits and harms should be reassessed at least every three months.

In treating chronic pain, opioids are an option to be tried after other less risky pain relievers have been considered, including paracetamol/acetaminophen or NSAIDs like ibuprofen or naproxen. Some types of chronic pain, including the pain caused by fibromyalgia or migraine, are preferentially treated with drugs other than opioids. The efficacy of using opioids to lessen chronic neuropathic pain is uncertain.

Opioids are contraindicated as a first-line treatment for headache because they impair alertness, bring risk of dependence, and increase the risk that episodic headaches will become chronic. Opioids can also cause heightened sensitivity to headache pain. When other treatments fail or are unavailable, opioids may be appropriate for treating headache if the patient can be monitored to prevent the development of chronic headache.

Opioids are being used more frequently in the management of non-malignant chronic pain. This practice has now led to a new and growing problem with addiction and misuse of opioids. Because of various negative effects the use of opioids for long term management of chronic pain is not indicated unless other less risky pain relievers have been found ineffective. Chronic pain which occurs only periodically, such as that from nerve pain, migraines, and fibromyalgia, frequently is better treated with medications other than opioids. Paracetamol and nonsteroidal anti-inflammatory drugs including ibuprofen and naproxen are considered safer alternatives. They are frequently used combined with opioids, such as paracetamol combined with oxycodone (Percocet) and ibuprofen combined with hydrocodone (Vicoprofen), which boosts the pain relief but is also intended to deter recreational use.

Other

Cough

Codeine was once viewed as the "gold standard" in cough suppressants, but this position is now questioned. Some recent placebo-controlled trials have found that it may be no better than a placebo for some causes including acute cough in children. Thus, it is not recommended for children. Additionally, there is no evidence that hydrocodone is useful in children. Similarly, a 2012 Dutch guideline regarding the treatment of acute cough does not recommend its use. (The opioid analogue dextromethorphan, long claimed to be as effective a cough suppressant as codeine, has similarly demonstrated little benefit in several recent studies.)

Low dose morphine may help chronic cough but its use is limited by side effects.

Diarrhea and constipation

In cases of diarrhea-predominate irritable bowel syndrome, opioids may be used to suppress diarrhea. Loperamide is a peripherally selective opioid available without a prescriptionused to suppress diarrhea.

The ability to suppress diarrhea also produces constipation when opioids are used beyond several weeks. Naloxegol, a peripherally-selective opioid antagonist is now available to treat opioid induced constipation.

Shortness of breath

Opioids may help with shortness of breath particularly in advanced diseases such as cancer and COPD among others.

ADVERSE EFFECTS

Adverse effects of opioids

Common and short term

- Itch
- Nausea
- Vomiting
- Constipation
- Drowsiness
- Dry mouth

Other

- Cognitive effects
- Opioid dependence
- Dizziness
- Decreased sex drive
- Impaired sexual function
- Decreased testosterone levels
- Depression
- Immunodeficiency
- Increased pain sensitivity
- Irregular menstruation
- Increased risk of falls
- Slowed breathing
- Coma

In older adults, opioid use is associated with increased adverse effects such as "sedation, nausea, vomiting, constipation, urinary retention, and falls". As a result, older adults taking opioids are at greater risk for injury. Opioids do not cause any specific organ toxicity, unlike many other drugs, such as aspirin and paracetamol. They are not associated with upper gastrointestinal bleeding and kidney toxicity.

Research suggests that when methadone is used long-term it can build up unpredictably in the body and lead to potentially deadly slowed breathing. Used medically, approaching toxicity goes unrecognized because the pain medication effect ends long before the drug's elimination half-life. According to the USCDC, methadone was involved in 31% of opioid related deaths in the US between 1999-2010 and 40% as the sole drug involved, far higher than other opioids. Studies of long term opioids have found that may stop them and minor side effects were common. Addiction occurred in about 0.3%. In the United States in 2016 opioid overdose resulted in the death of 1.7 in 10,000 people.

In the US charts below many deaths involve multiple opioids:

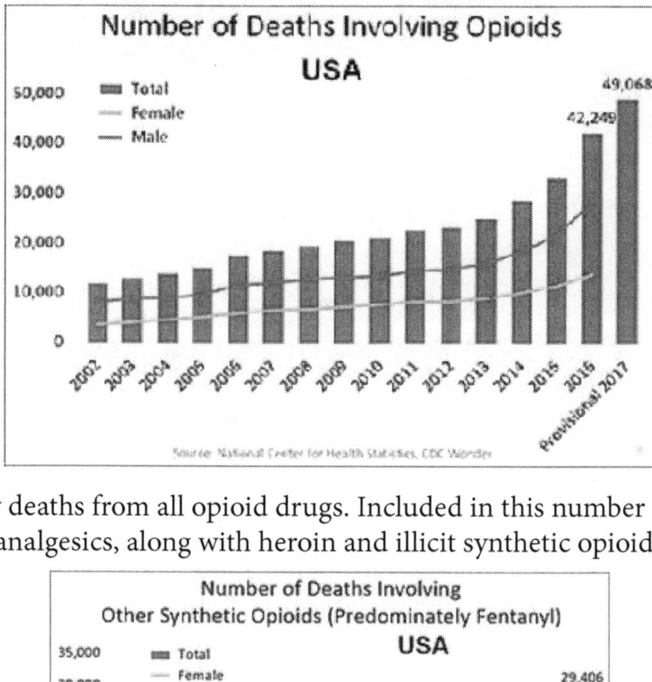

US yearly deaths from all opioid drugs. Included in this number are opioid analgesics, along with heroin and illicit synthetic opioids

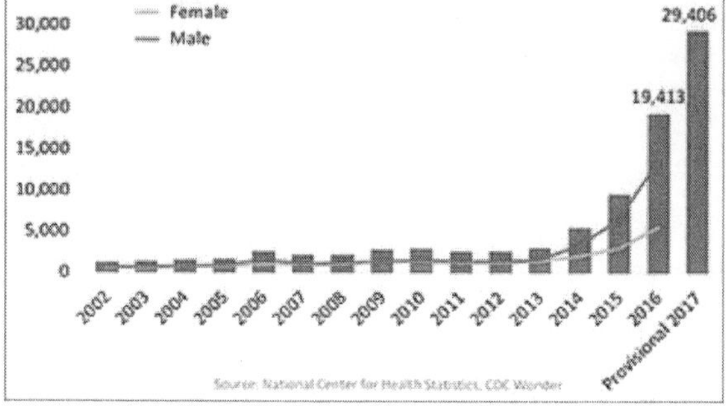

US yearly deaths involving other synthetic opioids, predominately Fentanyl.

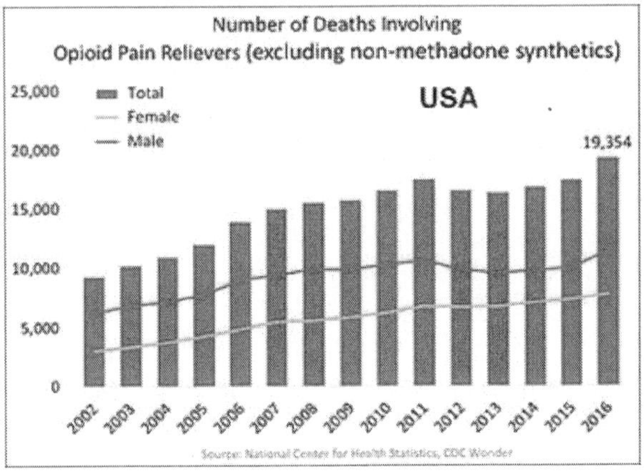

US yearly deaths involving prescription opioids. Non-methadone synthetics is a category dominated by illegally acquired fentanyl, and has been excluded.

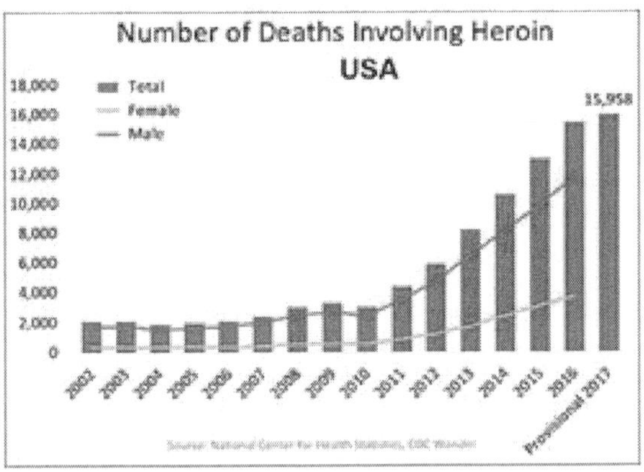

US yearly overdose deaths involving heroin

REINFORCEMENT DISORDERS

Tolerance

Tolerance is a process characterized by neuroadaptations that result in reduced drug effects. While receptor upregulation may often play an important role other mechanisms are also known. Tolerance is more pronounced for some effects than for others; tolerance occurs slowly to the effects on mood, itching, urinary retention, and respiratory depression, but occurs more quickly to the analgesia and other physical side effects. However, tolerance does not develop to constipation or miosis (the constriction of the pupil of the eye to less than or equal to two millimeters). This idea has been challenged, however, with some authors arguing that tolerance does develop to miosis.

Tolerance to opioids is attenuated by a number of substances, including:

- Calcium channel blockers
- Intrathecal magnesium and zinc
- NMDA antagonists, such as dextromethorphan, ketamine, and memantine.
- Cholecystokinin antagonists, such as proglumide
- Newer agents such as the phosphodiesterase inhibitor ibudilast have also been researched for this application.

Tolerance is a physiologic process where the body adjusts to a medication that is frequently present, usually requiring higher doses of the same medication over time to achieve the same effect. It is a common occurrence in individuals taking high doses of opioids for extended periods, but does not predict any relationship to misuse or addiction.

Physical dependence

Physical dependence is the physiological adaptation of the body to the presence of a substance, in this case opioid medication. It is defined by the development of withdrawal symptoms when the substance is discontinued, when the dose is reduced abruptly or, specifically in the case of opioids, when an antagonist (e.g., naloxone) or an agonist-antagonist (e.g., pentazocine) is administered. Physical dependence is a normal and expected aspect of certain medications and does not necessarily imply that the patient is addicted.

The withdrawal symptoms for opiates may include severe dysphoria, craving for another opiate dose, irritability, sweating, nausea, rhinorrea, tremor, vomiting and myalgia. Slowly reducing the intake of opioids over days and weeks can reduce or eliminate the withdrawal symptoms. The speed and severity of withdrawal depends on the half-life of the opioid; heroin and morphine withdrawal occur more quickly than methadone withdrawal. The acute withdrawal phase is often followed by a protracted phase of depression and insomnia that can last for months. The symptoms of opioid withdrawal can be treated with other medications, such as clonidine. Physical dependence does not predict drug misuse or true addiction, and is closely related to the same mechanism as tolerance. While there is anecdotal claims of benefit with ibogaine, data to support its use in substance dependence is poor.

Addiction

Drug addiction is a complex set of behaviors typically associated with misuse of certain drugs, developing over time and with higher drug dosages. Addiction includes psychological compulsion, to the extent that the sufferer persists in actions leading to dangerous or unhealthy outcomes. Opioid addiction includes insufflation or injection, rather than taking opioids orally as prescribed for medical reasons.

In European nations such as Austria, Bulgaria, and Slovakia, slow release oral morphine formulations are used in opiate substitution therapy (OST) for patients who do not well tolerate the side effects of buprenorphine or methadone. In other European countries including the UK, this is also legally used for OST although on a varying scale of acceptance.

Tamper-release formulations of time-controlled preparations of medications are intended to curb abuse and addiction rates while trying to still provide legitimate pain relief and ease of use to pain patients. Questions remain, however, about the efficacy and safety of these types of preparations. Further tamper resistant medications are currently under consideration with trials for market approval by the FDA.

The amount of evidence available only permits making a weak conclusion, but it suggests that a physician properly managing opioid use in patients with no history of substance dependence or substance abuse can give long-term pain relief with little risk of developing addiction, abuse, or other serious side effects.

Problems with opioids include the following:

1. Some people find that opioids do not relieve all of their pain.
2. Some people find that opioids side effects cause problems which outweigh the therapy's benefit
3. Some people build tolerance to opioids over time. This requires them to increase their drug dosage to maintain the benefit, and that in turn also increases the unwanted side effects.
4. Long-term opioid use can cause opioid-induced hyperalgesia, which is a condition in which the patient has increased sensitivity to pain.

All of the opioids can cause side effects. Common adverse reactions in patients taking opioids for pain relief include nausea and vomiting, drowsiness, itching, dry mouth, dizziness, and constipation.

Nausea and vomiting

Tolerance to nausea occurs within 7–10 days, during which antiemetics (e.g. low dose haloperidol once at night) are very effective. Due to severe side effects such as tardive dyskinesia, haloperidol is now rarely used. A related drug, prochlorperazine is more often used, although it has similar risks. Stronger antiemetics such as ondansetron or tropisetron are sometimes used when nausea is severe or continuous and disturbing, despite their greater cost. A less expensive alternative is dopamine antagonists such as domperidone and metoclopramide. Domperidone does not cross the blood–brain barrier and produce adverse central antidopaminergic effects, but blocks opioid emetic action in the chemoreceptor trigger zone. (The drug is not available in the U.S.) Some antihistamines with anticholinergic properties (e.g. orphenadrine or diphenhydramine) may also be effective. The first-generation antihistamine hydroxyzine is very commonly used, with the added advantages of not causing movement disorders, and also possessing analgesic-sparing properties. Ä-tetrahydrocannabinol relieves nausea and vomiting; it also produces analgesia that may allow lower doses of opioids with reduced nausea and vomiting.

- 5-HT3 antagonists (e.g. ondansetron)
- Dopamine antagonists (e.g. domperidone)
- Anti-cholinergic antihistamines (e.g. diphenhydramine)
- Ä-tetrahydrocannabinol (e.g. dronabinol)

Vomiting is due to gastric stasis (large volume vomiting, brief nausea relieved by vomiting, oesophageal reflux, epigastric fullness, early satiation), besides direct action on the chemoreceptor trigger zone of the area postrema, the vomiting centre of the brain. Vomiting can thus be prevented by prokinetic agents (e.g. domperidone or metoclopramide). If vomiting has already started, these drugs need to be administered by a non-oral route (e.g. subcutaneous for metoclopramide, rectally for domperidone).

- Prokinetic agents (e.g. domperidone)
- Anti-cholinergic agents (e.g. orphenadrine)

Drowsiness

Tolerance to drowsiness usually develops over 5–7 days, but if troublesome, switching to an alternative opioid often helps. Certain opioids such as fentanyl, morphine and diamorphine (heroin) tend to be particularly sedating, while others such as oxycodone, tilidine and meperidine (pethidine) tend to produce comparatively less sedation, but individual patients responses can vary markedly and some degree of trial and error may be needed to find the most suitable drug for a particular patient. Otherwise, treatment with CNSstimulants is generally effective.

- Stimulants (e.g. caffeine, modafinil, amphetamine, methylphenidate)

Itching

Itching tends not to be a severe problem when opioids are used for pain relief, but antihistamines are useful for counteracting itching when it occurs. Non-sedating antihistamines such as fexofenadine are often preferred as they avoid increasing opioid induced drowsiness. However, some sedating antihistamines such as orphenadrine can produce a synergistic pain relieving effect permitting smaller doses of opioids be used. Consequently, several opioid/antihistamine combination products have been marketed, such as Meprozine (meperidine/promethazine) and Diconal (dipipanone/cyclizine), and these may also reduce opioid induced nausea.

- Antihistamines (e.g. fexofenadine)

Constipation

Opioid-induced constipation (OIC) develops in 90 to 95% of people taking opioids long-term. Since tolerance to this problem does not develop readily, most people on long-term opioids need to take a laxative or enemas. While all opioids cause constipation, there are some differences between drugs, with studies suggesting tramadol, tapentadol, methadone and fentanyl may cause relatively less constipation, while with codeine, morphine, oxycodone or hydromorphone constipation may be comparatively more severe. Opioid rotation is commonly used to try and minimise the impact of constipation in long-term users.

Treatment

Treatment of OIC is successional and dependent on severity. The first mode of treatment is non-pharmacological, and includes lifestyle modifications like increasing dietary fiber, fluid intake (around 1.5 L (51 US fl oz) per day), and physical activity. If non-pharmacological measures are ineffective, laxatives, including stool softeners (e.g., docusate), bulk-forming laxatives (e.g., fiber supplements), stimulant laxatives (e.g., bisacodyl, senna), and/or enemas, may be used. A common laxative regimen for OIC is the combination of docusate and bisacodyl. Osmotic laxatives, including lactulose, polyethylene glycol, and milk of magnesia (magnesium hydroxide), as well as mineral oil (a lubricant laxative), are also commonly used for OIC.

If laxatives are insufficiently effective (which is often the case), opioid formulations or regimens that include a peripherally-selective opioid antagonist, such as methylnaltrexone bromide, naloxegol, alvimopan, or naloxone (as in oxycodone/naloxone), may be tried. A 2008 Cochrane review found that the evidence was tentative for alvimopan, naloxone, or methylnaltrexone bromide.

Respiratory depression

Respiratory depression is the most serious adverse reaction associated with opioid use, but it usually is seen with the use of a single, intravenous dose in an opioid-naïve patient. In patients taking opioids regularly for pain relief, tolerance to respiratory depression occurs rapidly, so that it is not a clinical problem. Several drugs have been developed which can partially block respiratory depression, although the only respiratory stimulant currently approved for this purpose is doxapram, which has only limited efficacy in this application. Newer drugs such as BIMU-8 and CX-546 may be much more effective.

- Respiratory stimulants: carotid chemoreceptor agonists (e.g. doxapram), 5-HT4 agonists (e.g. BIMU8), ä-opioid agonists (e.g. BW373U86) and AMPAkines (e.g. CX717) can all reduce respiratory depression caused by opioids without affecting analgesia, but most of these drugs are only moderately effective or have side effects which preclude use in humans. 5-HT1A agonists such as 8-OH-DPAT and repinotan also counteract opioid-induced respiratory depression, but at the same time reduce analgesia, which limits their usefulness for this application.
- Opioid antagonists (e.g. naloxone, nalmefene, diprenorphine)

Opioid-induced hyperalgesia

Opioid-induced hyperalgesia – where individuals using opioids to relieve pain paradoxically experience more pain as a result of that medication – has been observed in some people. This phenomenon, although uncommon, is seen in some people receiving palliative care, most often when dose is increased rapidly. If encountered, rotation between several different opioid pain medications may decrease the development of increased pain. Opioid induced hyperalgesia more commonly occurs with chronic use or brief high doses but some research suggests that it may also occur with very low doses.

Side effects such as hyperalgesia and allodynia, sometimes accompanied by a worsening of neuropathic pain, may be consequences of long-term treatment with opioid analgesics, especially when increasing tolerance has resulted in loss of efficacy and consequent progressive dose escalation over time. This appears to largely be a result of actions of opioid drugs at targets other than the three classic opioid receptors, including the nociceptin receptor, sigma receptor and Toll-like receptor 4, and can be counteracted in animal models by antagonists at these targets such as J-113,397, BD-1047 or (+)-naloxone respectively. No drugs are currently approved specifically for counteracting opioid-induced hyperalgesia in humans and in severe cases the only solution may be to discontinue use of opioid analgesics and replace them with non-opioid analgesic drugs. However, since individual sensitivity to the development of this side effect is highly dose dependent and may vary depending which opioid analgesic is used, many patients can avoid this side effect simply through dose reduction of the opioid drug (usually accompanied by the addition of a supplemental non-opioid analgesic), rotating between different opioid drugs, or by switching to a milder opioid with a mixed mode of action that also counteracts neuropathic pain, particularly tramadol or tapentadol.

- NMDA receptor antagonists such as ketamine
- SNRIs such as milnacipran
- Anticonvulsants such as gabapentin or pregabalin

OTHER ADVERSE EFFECTS

Low sex hormone levels

Clinical studies have consistently associated medical and recreational opioid use with hypogonadism (low sex hormone levels) in different sexes. The effect is dose-dependent. Most studies suggest that the majority (perhaps as much as 90%) of chronic opioid users suffer from hypogonadism. Opioids can also interfere with menstruation in women by limiting the production of luteinizing hormone (LH). Opioid-induced hypogonadism likely causes the strong association of opioid use with osteoporosis and bone fracture, due to deficiency in estradiol. It also may increase pain and thereby interfere with the intended clinical effect of opioid treatment. Opioid-induced hypogonadism is likely caused by their agonism of opioid receptors in the hypothalamus and the pituitary gland. One study found that the depressed testosterone levels of heroin addicts returned to normal within one month of abstinence, suggesting that the effect is readily reversible and is not permanent. As of 2013, the effect of low-dose or acute opioid use on the endocrine system is unclear.

Disruption of work

Use of opioids may be a risk factor for failing to return to work.

Persons performing any safety-sensitive task should not use opioids. Health care providers should not recommend that workers who drive or use heavy equipment including cranes or forklifts treat chronic or acute pain with opioids. Workplaces which manage workers who perform safety-sensitive operations should assign workers to less sensitive duties for so long as those workers are treated by their physician with opioids.

People who take opioids long term have increased likelihood of being unemployed. Taking opioids may further disrupt the patient's life and the adverse effects of opioids themselves can become a significant barrier to patients having an active life, gaining employment, and sustaining a career.

In addition, lack of employment may be a predictor of aberrant use of prescription opioids.

Increased accident-proneness

Opioid use may increase accident-proneness. Opioids may increase risk of traffic accidents and accidental falls.

Rare side effects

Infrequent adverse reactions in patients taking opioids for pain relief include: dose-related respiratory depression (especially with more potent opioids), confusion, hallucinations, delirium, urticaria, hypothermia, bradycardia/tachycardia, orthostatic hypotension, dizziness, headache, urinary retention, ureteric or biliary spasm, muscle rigidity, myoclonus (with high doses), and flushing (due to histamine release, except fentanyl and remifentanil). Both therapeutic and chronic use of opioids can compromise the function of the immune system. Opioids decrease the proliferation of macrophage progenitor cells and lymphocytes, and affect cell differentiation (Roy & Loh, 1996). Opioids may also inhibit leukocytemigration. However the relevance of this in the context of pain relief is not known.

Interactions

Physicians treating patients using opioids in combination with other drugs keep continual documentation that further treatment is indicated and remain aware of opportunities to adjust treatment if the patient's condition changes to merit less risky therapy.

With other depressant drugs

The concurrent use of opioids with other depressant drugs such as benzodiazepines or ethanol increases the rates of adverse events and overdose. As with an overdose of opioid alone, the combination of an opioid and another depressant may precipitate respiratory depression often leading to death. These risks are lessened with close monitoring by a physician, who may conduct ongoing screening for changes in patient behavior and treatment compliance.

US. Top line represents the number of benzodiazepine deaths that also involved opioids. Bottom line represents benzodiazepine deaths that did not involve opioids.

Opioid antagonist

Opioid effects (adverse or otherwise) can be reversed with an opioid antagonist such as naloxone or naltrexone.These competitive antagonists bind to the opioid receptors with higher affinity than agonists but do not activate the receptors. This displaces the agonist, attenuating or reversing the agonist effects. However, the elimination half-life of naloxone can be shorter than that of the opioid itself, so repeat dosing or continuous infusion may be required, or a longer acting antagonist such as nalmefene may be used. In patients taking opioids regularly it is essential that the opioid is only partially reversed to avoid a severe and distressing reaction of waking in excruciating pain. This is achieved by not giving a full dose but giving this in small doses until the respiratory rate has improved. An infusion is then started to keep the reversal at that level, while maintaining pain relief. Opioid antagonists remain the standard treatment for respiratory depression following opioid overdose, with naloxone being by far the most commonly used, although the longer acting antagonist nalmefene may be used for treating overdoses of long-acting opioids such as methadone, and diprenorphine is used for reversing the effects of extremely potent opioids used in veterinary medicine such as etorphine and carfentanil. However, since opioid antagonists also block the beneficial effects of opioid analgesics, they are generally useful only for treating overdose, with use of opioid antagonists alongside opioid analgesics to reduce side effects, requiring careful dose titration and often being poorly effective at doses low enough to allow analgesia to be maintained.

Pharmacology

Opioids bind to specific opioid receptors in the nervous system and other tissues. There are three principal classes of opioid receptors, ì, ê, ä (mu, kappa, and delta), although up to seventeen have been reported, and include the å, é, ë, and æ (Epsilon, Iota, Lambda and Zeta) receptors. Conversely, ó (Sigma) receptors are no longer considered to be opioid receptors because their activation is not reversed by the opioid inverse-agonist naloxone, they do not exhibit high-affinity binding for classical opioids, and they are stereoselective for dextro-rotatory isomers while the other opioid receptors are stereo-selective for levo-rotatory isomers. In addition, there are three subtypes of ì-receptor: ì1 and ì2, and the newly discovered ì3. Another receptor of clinical importance is the opioid-receptor-like receptor 1 (ORL1), which is involved in pain responses as well as having a major role in the development of tolerance to ì-opioid agonists used as analgesics. These are all G-protein coupled receptors acting on GABAergic neurotransmission.

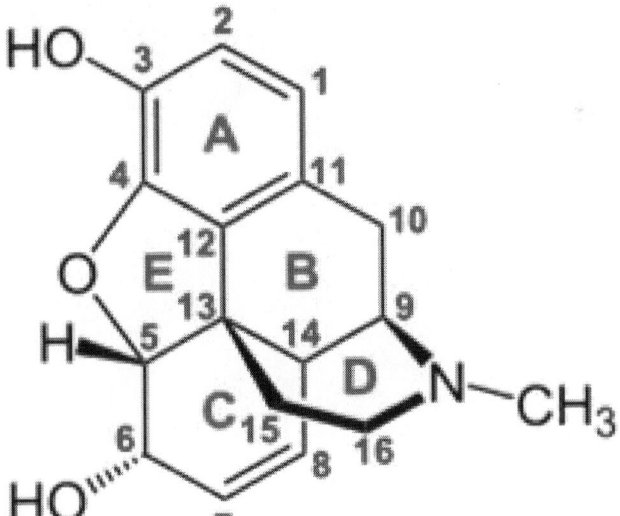

Locants of the morphine molecule

The pharmacodynamic response to an opioid depends upon the receptor to which it binds, its affinity for that receptor, and whether the opioid is an agonist or an antagonist. For example, the supraspinalanalgesic properties of the opioid agonist morphine are mediated by activation of the ì1 receptor; respiratory depression and physical dependence by the ì2 receptor; and sedation and spinal analgesia by the ê receptor. Each group of opioid receptors elicits a distinct set of neurological responses, with the receptor subtypes (such as ì1 and ì2 for example) providing even more [measurably] specific responses. Unique to each opioid is its distinct binding affinity to the various classes of opioid receptors (e.g. the ì, ê, and ä opioid receptors are activated at different magnitudes according to the specific receptor binding affinities of the opioid). For example, the opiate alkaloid morphine exhibits high-affinity binding to the ì-opioid receptor, while ketazocine exhibits high affinity to 8 receptors. It is this combinatorial mechanism that allows for such a wide class of opioids and molecular designs to exist, each with its own unique effect profile. Their individual molecular structure is also responsible for their different duration of action, whereby metabolic breakdown (such as N-dealkylation) is responsible for opioid metabolism.

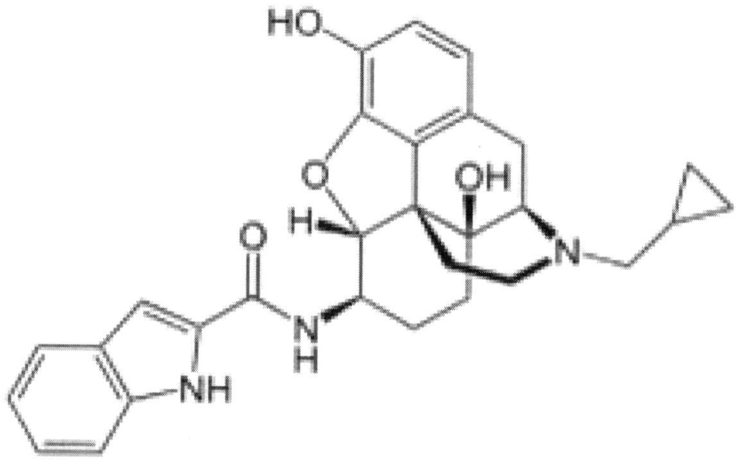

INTA: selective agonist of KOR-DOR and KOR-MOR heteromers. Does not recruit â-arrestin II. Antinociceptive devoid of aversion, tolerance, and dependence in mice.

Functional selectivity

A new strategy of drug development takes receptor signal transduction into consideration. This strategy strives to increase the activation of desirable signalling pathways while reducing the impact on undesirable pathways. This differential strategy has been given several names, including functional selectivity and biased agonism. The first opioid that was intentionally designed as a biased agonist and placed into clinical evaluation is the drug oliceridine. It displays analgesic activity and reduced adverse effects.

Opioid comparison

Extensive research has been conducted to determine equivalence ratios comparing the relative potency of opioids. Given a dose of an opioid, an equianalgesic table is used to find the equivalent dosage of another. Such tables are used in opioid rotation practices, and to describe an opioid by comparison to morphine, the reference opioid. Equianalgesic tables typically list drug half-lives, and sometimes equianalgesic doses of the same drug by means of administration, such as morphine: oral and intravenous.

Binding profiles

In the 1990s, opioid prescribing increased significantly. Once used almost exclusively for the treatment of acute pain or pain due to cancer, opioids are now prescribed liberally for people experiencing chronic pain. This has been accompanied by rising rates of accidental addiction and accidental overdoses leading to death. According to the International Narcotics Control Board, the United States and Canada lead the per capita consumption of prescription opioids. The number of opioid prescriptions per capita in the United States and Canada is double the consumption in the European Union, Australia, and New Zealand. Certain populations have been affected by the opioid addiction crisis more than others, including First World communities and low-income populations. Public health specialists say that this may result from unavailability or high cost of alternative methods for addressing chronic pain.

History

Naturally occurring opioids

A sample of raw opium

Opioids are among the world's oldest known drugs. The earliest known evidence of Papaver somniferum in a human archaeological site dates to the Neolithic period around 5,700-5,500 BC. Its seeds have been found at Cueva de los Murciélagos in the Iberian Peninsula and La Marmotta in the Italian Peninsula.

Use of the opium poppy for medical, recreational, and religious purposes can be traced to the 4th century BC, when ideograms on Sumerians clay tablets mention the use of "Hul Gil", a "plant of joy". Opium was known to the Egyptians, and is mentioned in the Ebers Papyrus as an ingredient in a mixture for the soothing of children, and for the treatment of breast abscesses.

Opium was also known to the Greeks. It was valued by Hippocrates (c. 460 – c. 370 BC) and his students for its sleep-inducing properties, and used for the treatment of pain. The Latin saying "Sedare dolorem opus divinum est", trans. "Alleviating pain is the work of the divine", has been variously ascribed to Hippocrates and to Galen of Pergamum. The medical use of opium is later discussed by Pedanius Dioscorides (c. 40 – 90 AD), a Greek physician serving in the Roman army, in his five-volume work, De Materia Medica.

During the Islamic Golden Age, the use of opium was discussed in detail by Avicenna (c. 980 – June 1037 AD) in The Canon of Medicine. The book's five volumes include information on opium's preparation, an array of physical effects, its use to treat a variety of illness, contraindications for its use, its potential danger as a poison and its potential for addiction. Avicenna discouraged opium's use except as a last resort, preferring to address the causes of pain rather than trying to minimize it with analgesics. Many of Avicenna's observations have been supported by modern medical research.

Exactly when opium became known in India and China is uncertain, but opium was mentioned in the Chinese medical work K'ai-pao-pen-tsdo (973 AD) By 1590 AD, opium poppies were a staple spring crop in the Subahs of Agra, Oudh and Allahabad in what is now Uttar Pradesh.

The physician Paracelsus (ca.1493–1541) is often credited with reintroducing opium into medical use in Western Europe, during the German Renaissance. He extolled opium's benefits for medical use. He also claimed to have an "arcanum", a pill which he called laudanum, that was superior to all others, particularly when death was to be cheated. ("Ich hab' ein Arcanum — heiss' ich Laudanum, ist über das Alles, wo es zum Tode reichen will.") Later writers have asserted that Paracelsus' recipe for laudanum contained opium, but its composition remains unknown.

Laudanum

The term laudanum was used generically for a useful medicine until the 17th century. After Thomas Sydenham introduced the first liquid tincture of opium, "laudanum" came to mean a mixture of both opium and alcohol. Sydenham's 1669 recipe for laudanum mixed opium with wine, saffron, clove and cinnamon. Sydenham's laudanum was used widely in both Europe and the Americas until the 20th century. Other popular medicines, based on opium, included Paregoric, a much milder liquid preparation for children; Black-drop, a stronger preparation; and Dover's powder.

The opium trade

Opium became a major colonial commodity, moving legally and illegally through trade networks involving India, the Portuguese, the Dutch, the British and China, among others. The British East India Company saw the opium trade as an investment opportunity in 1683 AD. In 1773 the Governor of Bengal established a monopoly on the production of Bengal opium, on behalf of the East India Company. The cultivation and manufacture of Indian opium was further centralized and controlled through a series of acts, between 1797 and 1949. The British balanced an economic deficit from the importation of Chinese tea by selling Indian opium which was smuggled into China in defiance of Chinese government bans. This led to the First (1839–1842) and Second Opium Wars (1856–1860) between China and Britain.

Morphine

In the 19th century, two major scientific advances were made that had far-reaching effects. Around 1804, German pharmacist Friedrich Sertürner isolated morphine from opium. He described its crystallization, structure, and pharmacological properties in a well-received paper in 1817. Morphine was the first alkaloid to be isolated from any medicinal plant, the beginning of modern scientific drug discovery.

The second advance, nearly fifty years later, was the refinement of the hypodermic needle by Alexander Wood and others. Development of a glass syringe with a subcutaneous needle made it possible to easily administer controlled measurable doses of a primary active compound.

Morphine was initially hailed as a wonder drug for its ability to ease pain. It could help people sleep, and had other useful side effects, including control of coughing and diarrhea. It was widely prescribed by doctors, and dispensed without restriction by pharmacists. During the American Civil War, opium and laudanum were used extensively to treat soldiers. It was also prescribed frequently for women, for menstrual pain and diseases of a "nervous character". At first it was assumed (wrongly) that this new method of application would not be addictive.

Codeine

Codeine was discovered in 1832 by Pierre Jean Robiquet. Robiquet was reviewing a method for morphine extraction, described by Scottish chemist William Gregory (1803-1838). Processing the residue left from Gregory's procedure, Robiquet isolated a crystalline substance from the other active components of opium. He wrote of his discovery: "Here is a new substance found in opium ... We know that morphine, which so far has been thought to be the only active principle of opium, does not account for all the effects and for a long time the physiologists are claiming that there is a gap that has to be filled." His discovery of the alkaloid led to the developmemt of a generation of antitussive and antidiarrheal medicines based on codeine.

Semisynthetic and synthetic opioids

Synthetic opioids were invented, and biological mechanisms for their actions discovered, in the 20th century. Scientists have searched for non-addictive forms of opioids, but have created stronger ones instead. In England Charles Romley Alder Wright developed hundreds of opiate compounds in his search for a nonaddictive opium derivative. In 1874 he became the first person to synthesize diamorphine (heroin), using a process called acetylation which involved boiling morphine with acetic anhydride for several hours. Heroin received little attention until it was independently synthesized by Felix Hoffmann (1868–1946), working for Heinrich Dreser (1860–1924) at Bayer Laboratories. Dreser brought the new drug to market as an analgesic and a cough treatment for tuberculosis, bronchitis, and asthma in 1898. Bayer ceased production in 1913, after heroin's addictive potential was recognized.

Several semi-synthetic opioids were developed in Germany in the 1910s. The first, oxymorphone, was synthesized from thebaine, an opioid alkaloid in opium poppies, in 1914.Next, Martin Freund and Edmund Speyer developed oxycodone, also from thebaine, at the University of Frankfurt in 1916. In 1920, hydrocodone was prepared by Carl Mannichand Helene Löwenheim, deriving it from codeine. In 1924, hydromorphone was synthesized by adding hydrogen to morphine. Etorphine was synthesized in 1960, from the oripavinein opium poppy straw. Buprenorphine was discovered in 1972.

The first fully synthetic opioid was meperidine (later demerol), found serendipitously by German chemist Otto Eisleb (or Eislib) at IG Farben in 1932. Meperidine was the first opiate to have a structure unrelated to morphine, but with opiate-like properties. Its analgesis effects were discovered by Otto Schaumann in 1939. Gustav Ehrhart and Max Bockmühl, also at IG Farben, built on the work of Eisleb and Schaumann. They developed "Hoechst 10820" (later methadone) around 1937. In 1959 the Belgian physician Paul Janssen developed fentanyl, a synthetic drug with 30 to 50 times the potency of heroin. Nearly 150 synthetic opioids are now known.

Criminalization and medical use

Non-clinical use of opium was criminalized in the United States by the Harrison Narcotics Tax Act of 1914, and by many other laws. The use of opioids was stigmatizied, and it was seen as a dangerous substance, to be prescribed only as a last resort for dying patients. The Controlled Substances Act of 1970 eventually relaxed the harshness of the Harrison Act.

In the United Kingdom the 1926 report of the Departmental Committee on Morphine and Heroin Addiction under the Chairmanship of the President of the Royal College of Physicians reasserted medical control and established the "British system" of control—which lasted until the 1960s.

In the 1980s the World Health Organization published guidelines for prescribing drugs, including opioids, for different levels of pain. In the U.S., Kathleen Foley and Russell Portenoy became leading advocates for the liberal use of opioids as painkillers for cases of "intractable non-malignant pain". With little or no scientific evidence to support their claims, industry scientists and advocates suggested that chronic pain sufferers would be resistant to addiction.

The release of OxyContin in 1996 was accompanied by an aggressive marketing campaign promoting the use of opioids for pain relief. Increasing prescription of opioids fueled a growing black market for heroin. Between 2000 and 2014 there was an "alarming increase in heroin use across the country and an epidemic of drug overdose deaths".

As a result, health care organizations and public health groups, such as Physicians for Responsible Opioid Prescribing, have called for decreases in the prescription of opioids. In 2016, the Centers for Disease Control and Prevention (CDC) issued a new set of guidelines for the prescription of opioids "for chronic pain outside of active cancer treatment, palliative care, and end-of-life care".

Society and culture

Definition

The term "opioid" originated in the 1950s. It combines "opium" + "-oid" meaning "opiate-like" ("opiates" being morphine and similar drugs derived from opium). The first scientific publication to use it, in 1963, included a footnote stating, "In this paper, the term, 'opioid', is used in the sense originally proposed by George H. Acheson (personal communication) to refer to any chemical compound with morphine-like activities". By the late 1960s, research found that opiate effects are mediated by activation of specific molecular receptors in the nervous system, which were termed "opioid receptors". The definition of "opioid" was later refined to refer to substances that have morphine-like activities that are mediated by the activation of opioid receptors. One modern pharmacology textbook states: "the term opioid applies to all agonists and antagonists with morphine-like activity, and also the naturally occurring and synthetic opioid peptides". Another pharmacology reference eliminates the morphine-like requirement: "Opioid, a more modern term, is used to designate all substances, both natural and synthetic, that bind to opioid receptors (including antagonists)". Some sources define the term opioid to exclude opiates, and others use opiatecomprehensively instead of opioid, but opioid used inclusively is considered modern, preferred and is in wide use.

Efforts to reduce abuse in the US

In 2011, the Obama administration released a white paper describing the administration's plan to deal with the opioid crisis. The administration's concerns about addiction and accidental overdosing have been echoed by numerous other medical and government advisory groups around the world.

As of 2015, prescription drug monitoring programs exist in every state, except for Missouri. These programs allow pharmacists and prescribers to access patients' prescription histories in order to identify suspicious use. However, a survey of US physicians published in 2015 found that only 53% of doctors used these programs, while 22% were not aware that the programs were available to them. The Centers for Disease Control and Prevention was tasked with establishing and publishing a new guideline, and was heavily lobbied. In 2016, the United States Centers for Disease Control and Prevention published its Guideline for Prescribing Opioids for Chronic Pain, recommending that opioids only be used when benefits for pain and function are expected to outweigh risks, and then used at the lowest effective dosage, with avoidance of concurrent opioid and benzodiazepine use whenever possible. Research has also suggested that automated algorithms may be able to identify patients who need more intensive screening and/ or intervention for non-medical opioid use; however, there is no standard for determining non-medical opioid use, and few algorithms have yet to be applied to real world settings.

GLOBAL SHORTAGES

Morphine and other poppy-based medicines have been identified by the World Health Organization as essential in the treatment of severe pain. As of 2002, seven countries (USA, UK, Italy, Australia, France, Spain and Japan) use 77% of the world's morphine supplies, leaving many emerging countries lacking in pain relief medication. The current system of supply of raw poppy materials to make poppy-based medicines is regulated by the International Narcotics Control Board under the provision of the 1961 Single Convention on Narcotic Drugs. The amount of raw poppy materials that each country can demand annually based on these provisions must correspond to an estimate of the country's needs taken from the national consumption within the preceding two years. In many countries, under prescription of morphine is rampant because of the high prices and the lack of training in the prescription of poppy-based drugs. The World Health Organization is now working with administrations from various countries to train healthworkers and to develop national regulations regarding drug prescription to facilitate a greater prescription of poppy-based medicines.

Another idea to increase morphine availability is proposed by the Senlis Council, who suggest, through their proposal for Afghan Morphine, that Afghanistan could provide cheap pain relief solutions to emerging countries as part of a second-tier system of supply that would complement the current INCB regulated system by maintaining the balance and closed system that it establishes while providing finished product morphine to those suffering from severe pain and unable to access poppy-based drugs under the current system.

Recreational use

Opioids can produce strong feelings of euphoria and are frequently used recreationally. Traditionally associated with illicit opioids such as heroin, prescription opioids are misused recreationally.

Drug misuse and non-medical use include the use of drugs for reasons or at doses other than prescribed. Opioid misuse can also include providing medications to persons for whom it was not prescribed. Such diversion may be treated as crimes, punishable by imprisonment in many countries. In 2014, almost 2 million Americans abused or were dependent on prescription opioids.

Classification

There are a number of broad classes of opioids:

- Natural opiates: alkaloids contained in the resin of the opium poppy, primarily morphine, codeine, and thebaine, but not papaverine and noscapine which have a different mechanism of action; The following could be considered natural opiates: The leaves from Mitragyna speciosa (also known as kratom) contain a few naturally-occurring opioids, active via Mu- and Delta receptors. Salvinorin A, found naturally in the Salvia divinorum plant, is a kappa-opioid receptor agonist.

- Esters of morphine opiates: slightly chemically altered but more natural than the semi-synthetics, as most are morphine prodrugs, diacetylmorphine (morphine diacetate; heroin), nicomorphine (morphine dinicotinate), dipropanoylmorphine (morphine dipropionate), desomorphine, acetylpropionylmorphine, dibenzoylmorphine, diacetyldihydromorphine;
- Semi-synthetic opioids: created from either the natural opiates or morphine esters, such as hydromorphone, hydrocodone, oxycodone, oxymorphone, ethylmorphine and buprenorphine;
- Fully synthetic opioids: such as fentanyl, pethidine, levorphanol, methadone, tramadol, tapentadol, and dextropropoxyphene;
- Endogenous opioid peptides, produced naturally in the body, such as endorphins, enkephalins, dynorphins, and endomorphins. Morphine, and some other opioids, which are produced in small amounts in the body, are included in this category.

Tramadol and tapentadol, which act as monoamine uptake inhibitors also act as mild and potent agonists (respectively) of the ì-opioid receptor. Both drugs produce analgesiaeven when naloxone, an opioid antagonist, is administered.

Some minor opium alkaloids and various substances with opioid action are also found elsewhere, including molecules present in kratom, Corydalis, and Salvia divinorum plants and some species of poppy aside from Papaver somniferum. There are also strains which produce copious amounts of thebaine, an important raw material for making many semi-synthetic and synthetic opioids. Of all of the more than 120 poppy species, only two produce morphine.

Amongst analgesics there are a small number of agents which act on the central nervous system but not on the opioid receptor system and therefore have none of the other (narcotic) qualities of opioids although they may produce euphoria by relieving pain—a euphoria that, because of the way it is produced, does not form the basis of habituation, physical dependence, or addiction. Foremost amongst these are nefopam, orphenadrine, and perhaps phenyltoloxamine or some other antihistamines. Tricyclic antidepressantshave painkilling effect as well, but they're thought to do so by indirectly activating the endogenous opioid system. Paracetamol is predominantly a centrally acting analgesic (non-narcotic) which mediates its effect by action on descending serotoninergic (5-hydroxy triptaminergic) pathways, to increase 5-HT release (which inhibits release of pain mediators). It also decreases cyclo-oxygenase activity. It has recently been discovered that most or all of the therapeutic efficacy of paracetamol is due to a metabolite, AM404, which enhances the release of serotonin and inhibits the uptake of anandamide.

BIBLIOGRAPHY

- American Association of Pharmaceutical Scientists. Introduction to Pharmaceutical Science. Retrieved 30 May 2016.
- American College of Clinical Pharmacy. The definition of clinical pharmacy. Pharmacotherapy 2008;28(6):816-817.
- American Society of Consultant Pharmacists, Frequently Asked Questions
- Board of Pharmacy Specialties, Current Specialties
- BPS Approves Ambulatory Care Designation; Explores New Specialties in Pain and Palliative Care, Critical Care and Pediatrics
- Burke JM, Miller WA, Spencer AP, et al. (2008). Clinical pharmacist competencies(PDF). Pharmacotherapy. 28 (6): 806–815. doi:10.1592/phco.28.6.806.
- Edward Kremers, Glenn Sonnedecker (1986). Kremers and Urdang's History of pharmacy. Amer. Inst. History of Pharmacy. p.17. ISBN 0931292174
- Hadzovic, S (1997). Pharmacy and the great contribution of Arab-Islamic science to its development. Medicinski Arhiv (in Croatian). 51 (1–2): 47–50. ISSN 0025-8083. OCLC 32564530. PMID 9324574.
- has photos
- Health Care Cost Drivers: Spotlight on Specialty Drugs (PDF). September 2014. Retrieved 27 October 2014.
- Hepler CD, Strand LM (1990). Opportunities and responsibilities in pharmaceutical care. Am J Hosp Pharm. 47 (3): 533–43. PMID 2316538.
- History of Pharmacy Web Pages - Sweden´s oldest pharmacies Archived 23 June 2011 at the Wayback Machine

- John K. Borchardt (2002). The Beginnings of Drug Therapy: Ancient Mesopotamian Medicine. Drug News & Perspectives. 15 (3): 187–192. doi:10.1358/dnp.2002.15.3.840015. ISSN 0214-0934. PMID 12677263.
- Knapp, KK; Okamoto, MP; Black, BL (2005). ASHP survey of ambulatory care pharmacy practice in health systems—2004. American Journal of Health-System Pharmacy. 62 (3): 274–84. PMID 15719585.
- Levey M. (1973), ' Early Arabic Pharmacology', E. J. Brill; Leiden.
- London Free Press Regional News Archive, Canada Internet Pharmacy Merged In $3.8 Million Deal Archived 26 April 2012 at the Wayback Machine
- NBCH Action Brief: Specialty Pharmacy. Specialty Pharmacy December, 2013. Accessed 27 October 2014.
- Pharmacist Certification and Course Requirements. Learn.org. 2013–2018. Retrieved 6 April 2018.
- Protecting Patients from Counterfeit and Other Substandard Drugs/ Supply Chain Threats
- Sharif Kaf al-Ghazal, Journal of the International Society for the History of Islamic Medicine, 2004 (3), pp. 3-9 .
- Specialty Pharmacy Certification Board
- Strand LM (1990). Pharmaceutical care and patient outcomes: notes on what it is we manage. Top Hosp Pharm Manage. 10 (2): 77–84. PMID 10128568.
- The History of Pharmacy. Eminent Scientists.
- Titsingh, Isaac. (1834) Annales des empereurs du japon, p. 434.
- Wild, D. Specialty Pharmacy Continuum Carving a Specialty Niche Within the ACO Model. Volume 1 (Summer Issue). August, 2012.
- World Health Organization. World Health Statistics 2011 - Table 6: Health workforce, infrastructure and essential medicines. Geneva, 2011. Accessed 21 July 2011.
- al-Ghazal, Sharif Kaf (October 2003). The valuable contributions of Al-Razi (Rhazes) in the history of pharmacy during the Middle Ages (PDF). Journal of the International Society for the History of Islamic Medicine. 2 (4): 9–11. ISSN 1303-667X. OCLC 54045642.

INDEX